# Making EHS
# an Integral Part
# of Process Design

# Making EHS
# an Integral Part
# of Process Design

**Arthur D. Little, Inc.**

Acorn Park
Cambridge, Massachusetts

**American Institute of Chemical Engineers**

3 Park Avenue
New York, New York 10016-5991

It is sincerely hoped that the information presented in this volume will lead to an even more impressive
safety record for the entire industry; however, the American Institute of Chemical Engineers, its consultants,
CWRT/CCPS Subcommittee members, their employers, their employers' officers and directors, and Arthur D.
Little, Inc. disclaim making or giving any warranties or representations, express or implied, including with
respect to fitness, intended purpose, use or merchantability and/or correctness or accuracy of the content of
the information presented in this document. As between (1) American Institute of Chemical Engineers, its
consultants, CWRT/CCPS Subcommittee members, their employers, their employers' officers and directors,
and Arthur D. Little, Inc. and (2) the user of this document, the user accepts any legal liability or responsibility whatsoever for the consequences of its use or misuse.

# Contents

# 3
# Life-Cycle Stages

# 4
# The MERITT Approach

# 5
# MERITT Tools

# 6
# Application of MERITT

# 7
# Implementation Guidance

## Appendix A
## Additional Tools with Applicability to MERITT

## Appendix B
## Overview of the INSET Tools and Their Aims

## Appendix C
## The Business Case for Managing Process Safety

## Index

# Preface

Chemical manufacturers are acutely aware of the rapidly changing business environment they face—increased global competition in commodity chemicals; ever increasing environmental, health, and safety regulations; more rigorous customer demands; extensive investor pressure; and increased focus on societal image. Capital productivity—the ratio of revenues to cost of assets—of the US chemical industry has been steadily declining (U.S. Chemical Industry Statistical Handbook, 1996), driven in part by the process modifications required to meet stricter environmental regulations that seldom provide a concomitant productivity increase. The development of the next generation of chemical manufacturing processes needed to address these challenges, as well as maintain competitiveness in mature business segments will require a broad, multidisciplinary rethinking of the traditional development and commercialization process and will require new tools and methodologies for various practitioners to employ. The new designs must result in plants that assure process and operator safety, the sustained health of workers and the community, and the viability of the environment. Furthermore, designs must be robust so that the plants are capable of meeting stringent customer demands for quantity and quality under conditions of process variability.

Traditionally, businesses have addressed regulatory and corporate requirements for the design of a chemical process through a compliance-based review of EHS issues, usually occurring well along in the design process, with each EHS discipline being reviewed separately. The proposed changes that result from these reviews almost always increase capital costs, with little perceived value.

It is now realized that early issue identification allows engineers and chemists along with EHS staff to combine their problem solving skills and thus arrive at more innovative and cost-effective solutions.

This realization has led to a more proactive consideration of EHS in business decisions and, aided by a number of EHS paradigms, has fostered determinative evaluations that focus, at least in part, on the more formative stages in the development of a new process, a process redesign, or a process upgrade. Notable among these are pollution prevention (P2), design for the environment (DfE), inherent safety (IS), green chemistry (GC), and green technology (GT).

This book presents an approach—termed MERITT (Maximizing EHS Returns by Integrating Tools and Talents)—for enhancing process development through better integration of environmental, health, and safety evaluations. It draws upon critical components of inherent safety, pollution prevention, green chemistry, and related paradigms through selective adoption and adaptation of their existing tools, skills, and knowledge resources. The holistic approach to considering the impacts and opportunities associated with the EHS characteristics of a given process, particularly during the early stages of the life-cycle yield better solutions that address the broad spectrum of EHS issues while also adding overall commercial benefits. Such solutions can also be realized more quickly than the less ideal solutions arrived at through sequential reviews.

MERITT has been formulated by drawing upon existing best practices of organizations that are recognized leaders in developing and conducting coordinated EHS evaluations through the use of tools and protocols devised specifically to support their own existing programs. MERITT offers ways of enhancing these best practices through increased awareness and recognition throughout all levels of the organization (engineers, chemists, project managers, and business leaders).

Integration of EHS encourages assessment of the full range of design opportunities, some of which may not be within the focus of any one individual paradigm. Integration leads to more effective decision making throughout the development cycle, especially in the early stages, and closer alignment of these decisions with business criteria translates directly into more viable processes, with faster commercialization and a lower cost.

Processes designed following the principles of the *individual* paradigms contained within MERITT have led to a broad array of cost savings, including increased process productivity, reduced waste generation, lower liabilities for fines and other penalties, greater optimization of levels of protection, and numerous other financial and business benefits. Implementation of MERITT may take slightly more resources up front in the process development cycle, but these resources will be more than offset by reduced resource requirements later in the development cycle along with significant operating cost savings.

# Acknowledgments

This book represents the collaborative efforts of a number of individuals and organizations. Arthur D. Little, Inc. was contracted to help develop and refine the project concepts and author the resulting book.

The Arthur D. Little staff included the following primary authors:

Lisa M. Bendixen
R. Peter Stickles
Richard R. Lunt

Peter W. Kopf and Maria P. Verzbolovskis provided additional support.

Information for the preparation and writing of this book was also obtained from the following organizations:

| | |
|---|---|
| 3M | ExxonMobil Chemical |
| AEA Technology | Loughborough University |
| AIChE | Monsanto |
| Air Products | NC State |
| Battelle—PNWL | Procter & Gamble |
| CCPS | Rohm and Haas |
| Dow | GlaxoSmithKline |
| DuPont | United Technologies Research Center |
| Equilon Environmental Directorate | University of Alabama |
| ExxonMobil | |

The Inherent Safety and Pollution Prevention subcommittee initiated this project in December of 1998. The members of this subcommittee are hereby recognized for their sustained efforts in this endeavor. They include the following:

Gregory L. Keeports, SUBCOMMITTEE CHAIR, *Rohm and Haas Company*
Dave Constable, *GlaxoSmithKline*

Leslie J. Cunningham III, *Merck & Co.*
Darryl W. Hertz, *Kellogg Brown & Root, Inc.*
Ronald C. Lutz, *Syngenta Crop Protection, Inc.*
Fred L. Maves, *3M*
Kenneth Mulholland, *Kenneth Mulholland & Associates, Inc.*
Jo Rogers, *AIChE–CWRT*
Jerry Schinaman, *Bristol-Myers Squibb Company*
Dan Sliva, *AIChE–CWRT*
Robert W. Sylvester, *DuPont Company*.

The subcommittee would also like to recognize the efforts of Manian Ramesh of Nalco Chemical Company and Stephen G. Maroldo, Rohm and Haas Company for their helpful comments, suggestions and guidance in their critique of the draft document. The support of additional reviewers at various stages of the project was also greatly appreciated.

Finally, the subcommittee would like to acknowledge the AIChE Foundation whose financial support made the preparation of this book possible.

# Acronyms

| | |
|---|---|
| AIChE | American Institute of Chemical Engineers |
| API | American Petroleum Institute |
| CCPS | Center for Chemical Process Safety |
| CERES | Coalition for Environmentally Responsible Economies |
| CIA | Chemical Industries Association |
| CWRT | Center for Waste Reduction Technologies |
| DfE | Design for the Environment |
| DFES | Design For Environment, Health, and Safety |
| E | Environmental |
| E/DS | Option Evaluation/Decision Support |
| EHS | Environment(al), Health, and Safety |
| EPA | Environmental Protection Agency |
| FDA | Food and Drug Administration |
| FMEA | Failure Modes and Effects Analysis |
| FN | Frequency-Number |
| GC | Green Chemistry |
| GCES | Green Chemistry Expert System |
| GRI | Global Reporting Initiative |
| GT | Green Technology |
| H | Health |
| HAZOP | Hazard and Operability (study) |
| I | Inquiry |
| INSET | Inherent SHE Evaluation Tool |
| INSIDE | INherent SHE In DEsign |
| IS | Inherent Safety |
| ISO | International Organization for Standardization |
| K-T | Kepner-Tregoe |
| LC50 | Lethal Concentration (50%) |

| LCA | Life-Cycle Analysis |
|-----|---------------------|
| LCI | Life-Cycle Inventory |
| LCS | Life-Cycle Stages |
| LD50 | Lethal Dose (50%) |
| LOPA | Layer of Protection Analysis |
| LTIR | Lost Time Incident Rate |
| M&EB | Mass and Energy Balance |
| MERITT | Maximizing EHS Returns by Integrating Tools and Talents |
| NESHAP | National Emission Standards for Hazardous Air Pollutants |
| NRDC | Natural Resources Defense Council |
| OEM | Original Equipment Manufacturer |
| OG | Option Generation |
| ORC | Organization Resource Counselors |
| OSHA | Occupational Safety and Health Administration |
| P&ID | Piping and Instrument Diagram |
| P2 | Pollution Prevention |
| PFD | Process Flow Diagram |
| PHA | Process Hazards Analysis |
| PPE | Personal Protective Equipment |
| PSM | Process Safety Management |
| QRA | Quantitative Risk Assessment |
| RCRA | Resource Conservation and Recovery Act |
| RMP | Risk Management Plan |
| ROI | Return On Investment |
| S | Safety |
| SG | Stage Gate |
| TCA | Total Cost Assessment |
| TCI | Total Cost Inventory |
| TLV | Threshold Limit Value |
| VOHAP | Volatile Organic Hazardous Air Pollutant |
| WBCSD | World Business Council on Sustainable Development |

# 1

# Introduction

This book presents an approach—termed MERITT (Maximizing EHS Returns by Integrating Tools and Talents)—for enhancing process development through more effective integration of environmental, health, and safety evaluations. MERITT has been based on the benchmarked best practices of industry leaders in this field and draws upon critical components of pollution prevention, inherent safety, green chemistry, and related paradigms through selective adoption and adaptation of their existing tools, skills, and knowledge resources. MERITT emphasizes concurrency of environmental, health, and safety thinking, resulting in a unified EHS perspective and its infusion into existing business processes. Rather than formulating a new, stand-alone methodology, this unified approach offers the flexibility to meet many different development needs, coupled with the capability to evolve with changing business imperatives.

This presentation of MERITT is but a first step in a continuing effort to realize the benefits to be obtained from advanced process development systems that integrate EHS perspectives. The book provides the reader with the basic philosophy and several suggested approaches and tools for successfully implementing MERITT, but it is not a detailed cookbook or tutorial. Nor can it be, given the need to integrate MERITT into each organization's own unique processes and culture.

## 1.1. The Need for MERITT

Traditionally, businesses have addressed regulatory and corporate requirements for the design of a chemical process through a compliance-based review of EHS issues, usually occurring well along in the design process, with each EHS discipline being reviewed separately. The

Experienced process/project professionals feel that there is commonly a 15–35% life-cycle cost reduction available when EHS issues are addressed in a concurrent and timely manner. Improvements of 50% or more in project costs have been attained in some cases.

proposed changes that result from these reviews almost always increase capital costs, with little perceived value. It is now realized that early issue identification allows engineers and chemists along with EHS staff to combine their problem solving skills and thus arrive at more innovative and cost-effective solutions. This realization has led to a more proactive consideration of EHS in business decisions and has been aided by a number of EHS paradigms that foster determinative evaluations that focus, at least in part, on the more formative stages in the development of a new process, a process redesign, or a process upgrade. Notable among these are pollution prevention (P2), design for the environment (DfE), inherent safety (IS), green chemistry (GC), and green technology (GT). These share many common themes and, in this regard, are somewhat overlapping (see Figure 1-1). Most importantly, though, the individual paradigms promote heightened awareness and encourage EHS evaluations at early stages of process development.

The more advanced paradigms have some form of methodology for identifying and addressing EHS issues, such as the strategies of Minimize, Moderate, Simplify, and Substitute as advocated in IS, or the structured brainstorming of P2 directed toward Substitution, Recycle, and Reuse. Several paradigms have formulated new tools to support assessments and decision making—such as the solvent selection guide developed under GC and the new Total Cost Assessment (TCA) tool for P2 (see Chapter 5). Most importantly, though, the individual paradigms promote heightened awareness and encourage EHS evaluations at the earliest stages of process development.

Nevertheless, no single paradigm has yet to attain broad-based support of multidisciplinary teams (chemists; engineers; environmental,

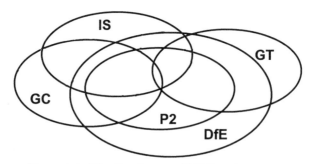

*Figure 1-1. The Overlapping EHS Paradigms.*

Industry is not fully exploiting the full range of opportunities available throughout the process development life cycle. There is enormous additional benefit to be harvested in integrating the concepts endorsed by P2, DfE, IS, GC, and GT within a programmatic effort that links information, decision making, and validated results throughout the development process.

health, and safety professionals; and business managers). Nor has any one paradigm gained preeminence as an overarching, all-encompassing framework. Instead, they remain separate and distinct, each tending to emphasize different aspects of the EHS whole (see Figure 1-2). Each has an appeal to an identifiable group of practitioners typically with different functions and accountabilities in process development, often in different industry segments and usually operating at a different stage in the life cycle or commercialization of a process. Therefore, the opportunities and benefits associated with each paradigm tend to be realized at different points in a project and in different ways. Also, because they are usually only partially overlapping, the areas of divergence necessitate judgment and reconciliation for which there is no current methodology. Consequently, few companies have fully embraced any of these paradigms in a systematic way by thoroughly infusing them in business processes that link all the steps in process or project development.

Integration and better timing lead to more effective decision making throughout the development cycle, especially in the early stages. The closer alignment of decisions and results with business criteria translates directly into more viable processes, with faster commercialization at lower costs.

| Paradigm | (Process) Safety | Health | Environment |
|---|---|---|---|
| Pollution Prevention (P2) | | ○ | ● |
| Design for the Environment (DfE) | | ○ | ● |
| Inherent Safety (IS) | ● | ● | ○ |
| Green Chemistry (GC) | ● | ○ | ● |
| Green Technology (GT) | ○ | | ● |

● Primary Focus          ● Secondary Focus          ○ Directly Linked Benefit

*Figure 1-2. Each paradigm has a distinct focus.*

The economies are significant in reducing iterative, parallel, and sequential evaluations and obviating the frequent add-on "fixes" that increase cost, delay schedules, and/or compromise performance. Integration also encourages assessment of the full range of opportunities, some of which may not be within the focus of any one paradigm. By the time they typically become involved, practitioners of E, H, or S often feel as if they have no choices when making recommendations on the changes needed to come into compliance. When the disciplines are considered earlier and together, there are many different ways in which an issue can be addressed because the design is not yet "set"—for example, you don't necessarily have to add a scrubber because you may be able to create a closed-loop process instead.

As a first step in developing this MERITT approach, a dozen or so major international petrochemical, chemical, and pharmaceutical companies were interviewed to better understand their practices and needs in this area. These companies are recognized leaders in developing and conducting coordinated E, H, and S evaluations through the use of tools and protocols developed specifically to support their own programs. In addition to identifying a number of specific programs followed by these companies, the interviews found that none of the programs entailed fully integrated EHS evaluations or full and early integration into the overall development process. Nonetheless, enough value had been shown in the limited efforts that were in place that there was interest in taking the idea of holistic, integrated evaluation further. However, no particular approaches were identified that could easily be adapted for use by others.

Further discussions were held, again with broad industry involvement, to collect further thoughts on:

- The preferred approach for integrating EHS into the development process
- Implementation suggestions and issues
- Staffing issues and concerns
- Tools—both existing and needed
- Success stories, as well as lessons from past failures and shortfalls
- General observations

Finding successful ways to work within existing processes and avoid creating new teams or additional demands on people's time were deemed necessary for future success by several participants. The discussions particularly served to point out concerns about perceived conflicts between EHS and process development efforts, as well as internal EHS conflicts that would need to be explicitly addressed. Strong biases in favor of existing programs applied by individual companies (such as P2) were also apparent. Creating a sense of awareness and enlightenment as well as sharing success stories and business cases to prove the value of the concept were widely thought by those participating to be the best means of overcoming these challenges.

MERITT, therefore, is a next, logical evolutionary step in the continuous improvement of existing process development. There are several drivers that compel this advancement.

- *Intensifying regulations*—While existing emissions limits continue to become more restrictive, there are new broad-based regulatory initiatives (e.g., for ozone nonattainment, fine particulate matter, $NO_x$, $CO_2$, and reactive chemicals) that will necessitate more "front-end" thinking for practical compliance.
- *Public advocacy*—Expectations concerning all aspects of business performance with regard to environment, health, and safety have progressed from "need-to-know" through "right-to-know" to mandated participation that now empowers a community with control over the company's "license to operate."
- *Financial realities*—Business sustainability increasingly demands more effective and comprehensive risk management, covering not only traditional concerns such as potential releases and environmental liabilities, but also sources of business interruption.
- *Corporate policy mandates*—Leadership in Responsible Care and Product Stewardship must be defensible and justifiable. Some companies have embraced concepts such as eco-efficiency, sustainable development, or the triple bottom line (which balances economic, social, and environmental factors) as their hallmark.
- *Global competitive pace*—Being the leader in any market increasingly means being the first to commercialize the most viable process in terms of cost, product and process quality, and EHS performance.

In attempting to deal with these pressures, current systems for process development are hampered by three structural deficiencies.

1. *Compartmentalism*—Compartmentalism exists in several forms in current process development systems. First, the process development system itself is usually compartmentalized with regard to moving from one stage to the next, with the stage gates (used for management of the process) serving as barriers to continuous information transfer. Second, in all but a very few companies, E, H, and S evaluations are still undertaken as separate and, at best, loosely coordinated activities. And, finally, EHS evaluations are usually separate from the mainstream of process development, both in timing and team interaction. While most of the necessary evaluations are still done, the norm is iterative or reactive reviews producing a succession of independently developed results. This can lead to "decisions by default" on the part of project leaders without the benefit of a full hearing on all issues or discussion of alternatives which can impose requirements that could have been avoided by early and continuing analysis of options.

2. *Overlapping exclusion*—The tendency is to embrace one or two paradigms as the cornerstone(s) of a company's environmental program—for example, GC or GT or P2. Often, the selection reflects the paradigm that is most closely aligned with the company's perceived business interests or what the recognized scientific leadership advocates. Expanding or adding more paradigms is thus viewed as overly complicating (or "muddying") an already crowded process. As shown in Figure 1-3, though, no single paradigm is truly comprehensive in addressing the full range of EHS issues. (Note that in Figure 1-3, the concepts are grouped within the context of the four basic IS strategies for convenience.) In fact, they endorse concepts derived from basic tenets that can collide in practice, as often happens in considerations of safety and environment. Examples include process simplification versus internal recycle (which is in itself complicating), and water minimization versus use of water as a diluent or heat sink to enhance process safety. Lacking an overarching perspective, it is almost impossible to address these potentially conflicting, but potentially useful concepts.

3. *"Schedulitis"*—Current systems for the management of process development preclude full integration of EHS opportunities. This

| Strategy/ Tenet (Based on IS) | Example Concepts | IS | P2 | GC | GT | DfE |
|---|---|---|---|---|---|---|
| Substitution | Reaction chemistry, Feedstocks, Catalysts, Solvents, Fuel selection | | | | | |
| Minimization | Process Intensification, **Recycle,** Inventory reduction, **Energy efficiency,** Plant location | | | | | |
| Simplification | Number of unit operations, DCS configuration, Raw material quality, Equipment design | | | | | |
| Moderation (1) [Basic Process] | Conversion conditions, Storage conditions, Dilution, Equipment overdesign | | | | | |
| Moderation (2) [Overall Plant] | **Offsite reuse,** Advanced waste treatment, Plant location, Beneficial co-disposal | | | | | |

Key:   ■ Primary tenet/concepts; ■ Strongly related tenet/concepts; ▨ Some aspects addressed; ☐ Little relationship.

*Figure 1-3. Paradigm comparison matrix.*

requires broader-based thinking earlier in the overall process and the creation of continuity through succeeding stages. Even for the conventional process development effort with routine stage gates, too often the hue and cry is that there isn't enough time. At the same time there is a growing need to be able to handle fast-tracked projects in a very growth-oriented global business environment. Programs founded in green chemistry and inherent safety are still in their infancy. And while these programs do focus on the early stages of development activities, they have yet to be widely embraced. None has yet to adequately address the needs of highly compressed, parallel processing of very fast- (or "flash-") tracked projects. Nonetheless, a well-coordinated approach should have a shorter schedule. The issues are getting the approach in place and tuning the process to the demands of the entire development schedule, not just the project schedule.

In summary, EHS evaluations are too frequently done serially, too late, and/or without adequately recognizing tradeoffs or resolving conflicting results. The development of MERITT is a step toward addressing these limitations—namely, moving considerations to the earliest stages in the process development life cycle and fostering simultaneous, coordinated thinking about EHS issues with other business imperatives. This inevitably leads to three extraordinary benefits:

- A higher probability of arriving at the *right answer* within the constraints of the established criteria;
- A greater degree of *sustainability* in the life cycle of the process (less need to "revisit" EHS issues later on); and
- A much *more economical overall development process*.

In the long run this can result only in making everyone's job easier, better understood, and more effective (for both the producer and the customer).

## 1.2. What Does MERITT Offer?

MERITT is an implementation-oriented approach for advancing the optimization of EHS evaluations within projects for:

- New process development,
- Process redesign,
- Process retrofit and upgrade, and
- Process optimization.

MERITT seeks to better coordinate EHS reviews and evaluations with existing business processes. This coordination includes accessing infor-

MERITT is founded on the following three key concepts and emphasizes a five "C" theme:

1. Driving the evaluations up front, to the earliest stage in the process;
2. Creating a continuum of thinking which links successive stages and evaluations; and
3. Promoting expanded information and knowledge exchange.

The underlying five "C" theme throughout is *concurrency*—concurrency through enhanced *collaboration* and more effective *communication*, which, in turn, require *commitment* and consistency of effort (*continuity*). Together, these form the five fundamental principles for EHS integration.

mation, identifying opportunities, applying analytical tools, and balancing tradeoffs . . . all directed toward achieving more enlightened, efficient, effective, and timely decision-making. Paramount to this end is integrating thinking throughout the decision-making process and, in doing so, spanning the process from early concept development through commercialization, and beyond. As such, MERITT presents a unifying approach to enable the creation of an overarching process.

MERITT draws upon existing best practices of organizations that are recognized leaders in developing and conducting coordinated EHS evaluations through the use of tools and protocols devised specifically to support their own programs built upon P2, DfE, IS, GC, and/or GT. It also recognizes that the focus of most research chemists is on molecule development, not process issues, and that the lack of familiarity with IS, P2, and the other paradigms can be overcome through training, tools, realized improvements, and access to EHS resources.

MERITT is not fundamentally new. Rather, MERITT can be most accurately described as creative adaptation—rethinking, reorganizing, and redirecting the many pieces that already exist and work individually, but which have yet to be fully integrated or coordinated. MERITT offers ways to enhance these processes through increased awareness and recognition throughout all levels of the organization (engineers, chemists, project managers, and business leaders). It emphasizes the need to think about the various concepts at the same time, to infuse appropriate tools at the various project stages, and to get the right type of knowledge at the right time. In the extreme, MERITT can also be used to frame a new process development structure in cases where it is desired to rebuild existing business processes or where processes do not exist, whether in the chemical-based process industries or nearly any other manufacturing company or operation.

The extent to which the concepts assembled within the construct of MERITT are utilized is left to the discretion of individual practitioners. This can range from the one extreme of wholesale acceptance of the entire framework to the other of selective use of specific tools or guidelines. If MERITT serves as nothing more than a stimulant for increasing awareness and thereby accelerating more coordinated thinking throughout the life of a project or process, then it has been a success, since a basic precept of MERITT is improving the exchange of information and knowledge through more effective planning and communication.

## 1.3. Whom Is MERITT for?

Because it is integral to the process development effort, MERITT should involve all of a company's business leadership who define the process structure; management staff who provide the guidance; and technical staff who support and interact daily within the process. These individuals generally fall into seven groups.

MERITT can make the work of these professionals more effective in two important ways: first, by producing a superior process with enhanced business value through integrated task management that reduces overall life-cycle cost; and second, by minimizing their required efforts not only during the development process but also over the entire life cycle.

- *Corporate Leadership*—Senior executives define the basis for the development process by establishing the policies and setting the priorities within which the businesses must operate, and by determining market positioning and timing. Their recognition of the value of optimizing EHS evaluations through integrated activities and commitment to this approach is vital to its success.
- *Business Manager*—The push–pull relationship of the business manager and the market ultimately determines the product and, thereby, process performance criteria. The business manager has overall responsibility for conformance to company policies in the development effort and coordinating the process development process with other business processes such as marketing. Consequently, the business manager usually defines specific stage gate criteria that the development must meet within the context of corporate guidelines; and must ultimately approve the successful completion at each stage.
- *Scientists and Engineers*—Development and design professionals are too often shackled by decisions at the handoffs. In essence, "what has been agreed" will generally dictate "what will be." They need to get involved as early as possible in prior development efforts to impart knowledge and provide guidance as to the consequences of decisions along the way.

- *Process Development Leaders*—The process development leader is ultimately responsible for coordinating the development activities and "making the calls" in the day-to-day management of the process evolution through basic engineering. As the primary decision maker, the process development leader must often referee the tradeoffs of conflicting options and agendas. Only the most knowledgeable and enlightened process development leaders can hope to make the majority of these calls correctly and quickly without the benefit of coordinated evaluations.
- *Project (Engineering/Design/Construction) Managers*—The project manager is responsible for executing the project, usually from the start of Basic Engineering through plant/process startup and commissioning. The project manager typically inherits the basic process design concepts and must deal with the oversights and "add-ons" that accrue from decisions and assumptions made in earlier stages.
- *Plant/Operations Manager*—In addition to the obvious responsibilities for operating and maintaining the process, the plant/operations manager is also implicitly responsible for implementing necessary changes to improve the efficacy of the process and ensure that its performance meets the expectations of the business. Most process upgrades and retrofits fall within his or her purview.
- *EHS Practitioners*—These are the knowledge managers (coordinators) or stewards chartered with providing the guidance for day-to-day fulfillment of the company's policies and procedures for ensuring performance and protecting the enterprise. Their activities are often carried out in parallel or sequential efforts that preclude cross-fertilization and "feed-forward" loops.

## 1.4. About the Book

### 1.4.1. Scope

This book addresses the following aspects of MERITT:

- The value to be gained by optimizing process development through integrating E, H, and S activities within process development activities.

- Examples and descriptions of how EHS integration into the process development process can be accomplished by utilizing the basic structure and concepts of MERITT, along with guidelines for its implementation in several different scenarios.
- The types of tools, expertise, and organization integral to MERITT to support its implementation.

Details of the many available tools are not provided, nor are there exhaustive discussions of the dynamics of implementing MERITT. The tools have been defined in the references provided. The dynamics of implementation will be highly variable from one enterprise to another and even within different project types.

## 1.4.2. Organization

The book is organized into seven chapters, including this Introduction.

*Chapter 2: Value and Benefits*—The motivation for MERITT is discussed. Vignettes and anecdotal information are provided illustrating both the value to be derived from optimizing EHS evaluations and the occasional perils of the more traditional approach to EHS reviews.

*Chapter 3: Life-Cycle Stages*—The basic stages of typical process development efforts are delineated for new processes, process modifications and process upgrades. Stage constraints, resource requirements, information needs, and opportunities are discussed to set the context for Chapter 4.

*Chapter 4: The MERITT Approach*—The overall approach of MERITT is reviewed along with its basic components. Application to process development stages is discussed including opportunities for value creation.

*Chapter 5: MERITT Tools*—The role of tools and the types of tools available for MERITT are described including both "integrated" tools and "nonintegrated" tools. Nonintegrated tools are relatively general and deal with inquiry, option definition, and evaluation. Integrated tools incorporate a more comprehensive methodology to optimize EHS evaluations.

*Chapter 6: Application of MERITT*—Two examples are discussed: process development for a new site when there is an existing process design to replicate or modify and the development of a new process involving a solvent substitution.

*Chapter 7: Implementation Guidance*—Guidance is provided as to how MERITT can be implemented within the context of existing management systems. Specific attention is paid to overcoming cultural barriers, variations among different industry sectors and application to small companies. Metrics are also discussed.

Appendices provide additional details on selected tools and diagrams for using MERITT at different project stages. Acronyms are also included.

## 1.5. The Path Forward

This book is not intended to be the definitive work on optimizing process development through full integration of E, H, and S evaluations. It is but another step in the ongoing attempt to advance the process of EHS integration. In helping to coordinate this effort, two of the American Institute of Chemical Engineers' Industry Technology Alliances, the Center for Waste Reduction Technologies (CWRT) and the Center for Chemical Process Safety (CCPS), have jointly established a program that will continue well beyond this book and will undoubtedly lead to additional publications and transmittal of information through other media.

AIChE's program comprises several interrelated activities, including:

- Disseminating information about MERITT and obtaining feedback through workshops and seminars;
- Developing guidance documents;
- Supporting trial applications and publishing case studies;
- Undertaking additional efforts to refine the approach; and
- Developing training materials.

To provide ongoing information about the program, AIChE has devoted a portion of its web site [aiche.org/cwrt] to MERITT and the program. The web site also contains a set of links to other useful sites related to optimizing EHS evaluations and furthering tool development.

# 2

# Value and Benefits

Chemical manufacturers are acutely aware of the rapidly changing business environment they face: increased global competition in commodity chemicals; ever increasing environmental, health, and safety regulations; more rigorous customer demands; extensive investor pressure; and increased focus on societal image. Capital productivity, the ratio of revenues to cost of assets, of the U.S. chemical industry has been steadily declining (*U.S. Chemical Industry Statistical Handbook,* 1996), driven in part by the process modifications required to meet stricter environmental regulations that seldom provide a concomitant productivity increase. The development of the next generation of chemical manufacturing processes needed to address these challenges, as well as maintain competitiveness in mature business segments, will require a broad, multidisciplinary rethinking of the traditional development and commercialization process and will require new tools and methodologies for various practitioners to employ. The new designs must result in plants that assure process and operator safety, the sustained health of workers and the community, and the viability of the environment. Furthermore, designs must be robust so that the plants are capable of making the required amount of product while meeting stringent customer demands for quality under conditions of process variability.

Through the combined practice of the paradigms of IS, GC, P2, DfE, and GT, companies of any size can realize increased benefits, such as reduced capital investments, over those available through an independent adherence to these paradigms. This chapter discusses the individual benefits offered by each approach as well as the greater benefits that can be gained through integration. Several examples are also provided, demonstrating what can all too easily happen when the concepts are not implemented in an integrated fashion.

## 2.1. The Value of MERITT

There are well-recognized benefits from following the individual paradigms of IS, P2, GC, DfE, or GT. Table 2-1 briefly summarizes a range of examples. Because so many of these paradigms and their techniques

**TABLE 2-1.** *Individual Value of Various MERITT-Related Paradigms*

| Paradigm | Example of Value to Business |
|---|---|
| **Inherent Safety** | The application of IS concepts has helped many companies avoid coming under process safety management (PSM) or other requirements through changes in the materials handled. For instance, a change in water treatment eliminated chlorine from many sites—often where it was the only high hazard material on the site—and substituted hypochlorite, which had the added benefit of eliminating the need to transport, store, and handle chlorine. |
| **Pollution Prevention** | Dow undertook a pollution prevention exercise at its Midland site in 1996, along with the Natural Resources Defense Council (NRDC) and various community activists. One of the processes that was modified was the manufacture of chloroacetylchloride (CAC). After reviewing options for improved cooling, optimization of the distillation column, and captive recycling, the selected approach was targeted at improved cooling via refrigeration. Keeping the reaction temperature lower both decreases the production of undesirable byproducts and increases the yield of CAC. The total reduction of byproduct wastes is estimated at 1.8 million pounds per year, and the project economics are such that the project more than breaks even in the first year. |
| | The overall pollution prevention exercise is expected to yield savings of $5.4 million annually, all for a total capital investment of $3.1 million. |
| **Green Chemistry** | BHC Co. (a joint venture of Celanese and BASF) received an EPA Presidential Green Chemistry Challenge Award in 1997 for its new three-step synthesis of ibuprofen. The first step minimizes waste as it allows for recycling with greater than 99.9% efficiency. The second and third steps are both solvent free and have efficiencies of 98%. This process replaces a six-step process, and is reputed to be the most cost-effective route for making ibuprofen while producing the highest quality product. |
| | The EPA has supported Syracuse Research Corporation's development of the Green Chemistry Expert System (GCES) to help further the application of GC to prevent pollution. |
| **Design for Environment** | United Technology Corporation touts significant performance gains in P2, waste minimization, and workplace safety through its Design for Environment, Health, and Safety (DFES) process, which fits within its integrated product development process. DFES has led to the use of simplified tools and approaches that UTC finds to be very efficient in terms of resources. |
| **Green Technology** | Releases of unburned fuel drove a green technology change for watercraft from 2-stroke engines (mixed oil and gas) to 4-stroke engines (gas only). As a result of this change, emissions to air and water were reduced from 25% to less than 5%. |

have demonstrated their cost savings many times over, it is rather surprising that more organizations have not adopted them, individually or collectively. The transfer of learning is easier for some of the paradigms, such as IS and P2, than for others, such as GC or DfE.

More in-depth examples of the individual and collective value of these paradigms are given below. However, most chemists involved in the early stages of research programs do not have formal training in IS, P2, GC, etc., and most engineers do not receive training in life-cycle analysis. Thus, there is significant opportunity for MERITT to add value beyond the individual paradigms as practiced today.

## 2.1.1. Making More Efficient Use of Resources

Thorough implementation of MERITT may take slightly more resources up front in the process development cycle, but these resources will be more than offset by reduced resource requirements later in the development cycle along with significant operating cost savings. In addition, the expenditure of additional resources in the early stages of process development can help ensure that MERITT does not have any negative impact on the overall schedule. The value of metrics, proving that MERITT really adds value, is discussed later in this chapter.

When a process is designed following the principles of inherent safety and pollution prevention, one expects a broad array of cost savings, particularly on the staffing and administrative side.

- Reduced need for initial and ongoing training, and therefore fewer trainers, both for a given individual and in terms of the total number of people requiring training. This results from fewer hazardous operations as well as less severe conditions where hazards remain, along with fewer regulatory requirements due to the less hazardous nature of the process.
- These same changes in the process will result in fewer personal protective equipment (PPE) requirements, leading to equipment and further training cost savings.
- Simplified and/or fewer safety procedures will lead to fewer opportunities for errors—as well as less time spent in maintaining and updating procedures.

- Less physical area and equipment needed for waste handling and storage, along with fewer requirements for and less time spent on monitoring, recordkeeping, etc.
- Lower manufacturing investments through elimination of high volume waste streams.
- Lower waste disposal costs.
- Reduced regulatory attention and even applicability or coverage in some situations.
- Lower equipment maintenance and operating costs.
- Decreased likelihood and/or consequences of releases or other types of failures.

As shown earlier in Table 2-1, some of these cost savings can be realized very quickly. Furthermore, as illustrated in the box below, the financial returns can be significant.

## 2.1.2. Examples of Success and Synergy

The individual paradigms contained within MERITT can and have led to increased process productivity, reduced waste generation, lower liabili-

A toxic surfactant was removed from a gas stream with a water scrubber and then removed from the water stream by means of ion exchange. The regenerate from the ion exchange bed was further concentrated using a reverse osmosis membrane process. The concentrated stream with the surfactant was then returned to the supplier for further purification. Over time, the ion exchange resin capacity was reduced and significant amounts of the expensive surfactant were lost.

To further reduce emissions of the surfactant, the collection system was expanded to include gas streams from other process units. A review of the recovery process showed that the ion exchange system, and its attendant costs and wastes, was not required to purify the water stream. The ion exchange system was replaced with a new reverse osmosis system to concentrate the surfactant-containing stream from the water scrubber to 1,000 ppm by weight. This 1,000 ppm by weight stream is then further concentrated to 20% by weight in the existing reverse osmosis system and the recovered water can then be recycled back to the water scrubbing system.

**The 140,000 lb/yr of recovered surfactant is returned to the supplier at a credit of $20/lb or $2,800,000/yr.**

ties for fines and other penalties, greater optimization of levels of protection, and numerous other financial and business benefits. While these benefits have largely been received as a result of following the paradigms individually, there are some examples where integrated approaches have been very rewarding. Formalizing integrated thinking should add value in many settings.

### Increasing Process Productivity

Ken Mulholland presents a case study in *Pollution Prevention: Methodology, Technologies, and Practices* (DuPont, 1999) that demonstrates the production efficiencies to be obtained from structured reviews of existing manufacturing processes. The example involves a business decision to relocate a process to a new site requiring more stringent requirements for environmental performance. The process involves air oxidation of reactants in the presence of steam and ammonia in a fixed-bed catalytic reactor. Water soluble product is then removed in an aqueous scrubber followed by benzene extraction with recovery of the benzene for reuse. Emissions abatement systems included an acid gas scrubber and a steam stripping column (producing wastewater) followed by a thermal oxidizer.

The investment associated with need for greatly enhanced end-of-pipe treatment of gaseous emissions and wastewater effluent spurred review of the process. The goals of the process review included lowering capital investments, reducing operating costs, eliminating the need to handle benzene, and reducing wastes.

Screening of the identified options to improve process design resulted in a dozen opportunities involving low capital investment, high capital investment, or additional R&D. These fell into four general categories:

1. Adjustments to the chemical synthesis route (e.g., use of steam with oxygen in place of air, a new catalyst, and elimination of the use of benzene);
2. Changes to equipment (e.g., use of a fluid-bed reactor and freeze crystallization technology);
3. Improved recovery/recycle of several streams; and
4. Water conservation.

**The revised process as finally approved resulted in a savings in capital investment of nearly $8 million with a total 10-year net present value savings of roughly $13 million.**

*Reducing Waste*

*Example 1*

The potential for major reductions in waste generation is illustrated well by Monsanto's "zero-waste" chemical process to make di-sodium iminodiacetate (DSIDA), which is a key intermediate in the production of Roundup® herbicide. This material cannot be purchased; it must be manufactured. Monsanto developed a new process in 1991 and had an operating plant in place by June of 1992.

Previously, there was a multistep process that used hydrogen cyanide, ammonia, and formalin. It also required the use of deep wells for the wastes. The cyanide was obtained from Chocolate Bayou (full production) and DuPont. Subject to both an increased demand for glyphosate (Roundup's active ingredient) and the manufacturer's concerns about selling cyanide even in lecture bottle quantities, there was a need to develop a new process.

The new process has only four steps, including the manufacture of DSIDA and a DSIDA "gobbling machine." It produces no waste and creates a combustible liquid. The process also creates huge quantities of hydrogen, which can be flared, blended with natural gas for boilers, or sold. Some of the lessons learned during the development of the new process include:

- Faster reacting (hotter) and more selective catalysts are better, saving on centrifuges, filters, and wasted material.
- Many batch chemistries can be converted to continuous processing.
- Major cost savings are available.
- Quicker times to market are possible.
- Additional benefits may still be available—for instance Monsanto is now using a similar catalyst in the last stage of the glyphosate process and is eliminating some of the formalin production.

*Example 2*

Novartis Crop Protection AG (formerly Ciba-Geigy AG) operates a state-of-the-art process and formulations development center in Muench-wilen, Switzerland. This center deals with process development for new agricultural chemical intermediates and active ingredients, formulation development, and piloting and small scale manufacturing. In the mid

1970s the following solvents were used during the development/scale-up of typical agricultural products and intermediates:

| | |
|---|---|
| Acetone | Acetonitrile |
| Acetic acid | Ethanol |
| Benzene | Nitrobenzene |
| Butanol | Monochlorobenzene |
| 1,2- 1,3- and 1,4-Di-chlorobenzenes | Dimethylformamide |
| Ethyl acetate | Butyl acetate |
| Cyclohexane | Cyclohexanol |
| Isopropanol | Methanol |
| 1,1- and 1,2-Di-chloroethane | Methylene chloride |
| Pyridine | Tetrahydrofuran |
| Chloroform | Toluene |
| Para-, Meta-, and Ortho-xylenes | 1,4-Dioxane |
| Methylethylketone | Methylisobutylketone |
| Methylisopropylketone | Dimethyl sulfoxide |
| Carbon disulfide | n-Hexane |
| n-Heptane | |

Through awareness that the selection of a solvent at an early stage can impact the overall economics of a process, the above list of chemicals was evaluated against certain criteria. These criteria included:

1. Water treatment implications for trace levels of the solvent
2. Air toxics classification
3. Industrial hygiene implications (carcinogenicity, TLVs, toxicity, electrostatic properties, etc.)
4. Vapor pressure
5. Boiling point
6. Halogenation (prefer nonhalogenated)
7. Energy requirements for evaporation
8. Ability to be recycled back to the same or similar chemical process

Through the past 20 years or so, the following "preferred" solvents were selected for emphasis for process development: methanol, acetone, ethanol, toluene, xylenes, isopropanol, dimethylformamide or n-methyl-pyrilidone, and cyclohexane.

Tank farm capacity formerly dedicated to handling a large variety of solvents was reconfigured to handle larger volumes of fewer solvents. This has also simplified logistics, and allowed rationalized deliveries of materials in tank trailers, rather than 200-liter drums. In the pilot facility the estimated savings are on the order of $33,000 per year.

### Controlling Legal Liabilities and Fines/Reducing Risks

An international food and consumer products corporation established an initiative to integrate its EHS evaluations several years ago. The initiative began with a focus on OSHA 1910 chemicals followed by those covered by EPA's Risk Management Plan (RMP), and lastly a prioritization of situations that had a combination of undesired chemical and/or process conditions. The prioritization considered the company's own incident tracking system as well as industry incidents with the same or similar chemicals. The latter grouping included: flammables ("Class I" chemicals); other high vapor pressure, hazardous chemicals; "Class IIIB" chemicals heated above boiling points; boilers and pressure vessels; process heaters; combustion systems; and process conditions representing special hazards—such as handling dusts. The integrated evaluations encompassed the products themselves, site operations, and transportation-related risks.

### Oleum Replacement

The company used oleum, typically brought in by rail, at a number of its facilities. Although the company had an excellent record in handling oleum, a release from an oleum rail car at another company's plant led to a reevaluation of the risks and benefits of continued use. Alternatives to the use of oleum were examined, as were enhanced methods for controlling the risks associated with the delivery, handling, storage, and use of oleum. Product efficacy was also a crucial element in these assessments. It was ultimately concluded that the optimal business solution was to produce oleum on-site via the oxidation of sulfur. While there are certainly risks associated with liquid sulfur and sulfur burning for production, these risks are generally far outweighed by the elimination of both the storage and transportation of oleum. Furthermore, the on-demand production of oleum confined most risks to potential on-site impacts and removed the risk of public exposure.

### Flammable Solvent

The integrated EHS initiative not only addressed existing production operations, but also new products and process development efforts. The success of the initiative was demonstrated in reviewing one of the new food products that had been in the R&D stage. It was determined through the integrated evaluation process that both safety and environmental risks could be significantly reduced by replacement of a relatively flammable solvent with one of lower vapor pressure. This early and simple process modification was far more cost effective to institute than it would have been if it were identified later in the product/process development cycle.

### Optimizing Layers of Protection

In this example, a traditional optimization of the layers of protection offered benefits, but not the same level of benefits as could be gained by a more fundamental change in the raw material. This illustrates the value that can be found in upgrades to existing processes.

For many years, a multinational specialty chemicals company had used large quantities of WXYZ as a feedstock for its manufacturing operations. WXYZ has a substantial vapor pressure and generates concentrated hydrochloric acid vapors when contacted with either liquid or vaporized water, a reaction that is highly exothermic. Thus, it is an extremely hazardous chemical with potential for significant health, safety, and environmental impacts not only due to its high corrosivity, but also to its potential for causing steam explosions as well as releases of HCl gas. In the manufacturing process, WXYZ was diluted with water prior to use, greatly reducing its hazardous characteristics (rendering it more like concentrated hydrochloric acid).

The company had rigorous, formal procedures and safeguards in handling WXYZ; nevertheless there were occasional minor incidents and near misses. Finally, in response to one incident, the company decided to rethink its operations to eliminate (or at least greatly minimize) future risks. Parallel evaluations were undertaken to investigate three general approaches:

1. Production alternatives for entirely eliminating the use of WXYZ;
2. Better ways to receive, store, and handle/use WXYZ on site; and
3. Options for preparing diluted WXYZ prior to delivery to the plant.

No production alternatives were found to be feasible. Maintaining product quality would require completely "retooling" the production facility, a very expensive proposition that had two very important business risks. First, the possibility that equivalent quality product could not be made, or made as efficiently and reproducibly. And second, the concern that the loss of production during changeover as well as production capacity decreases could cost the company valuable customers. The only improvements to handling WXYZ on site were determined to involve additional layers of protection, covering administrative and operational procedures as well as engineered interlocks and monitoring systems. Most were capable of reducing the frequency of occurrences, but not the magnitude of impacts.

In contrast to these two approaches, dilution of WXYZ at the suppliers' facilities prior to transport held enormous benefits of permanent risk reduction in all areas—safety, health, and environment. Therefore, the company, in cooperation with each of its suppliers, developed a design for the dilution system to be installed at the suppliers' facilities. These "partnerships" established cost-sharing agreements not only for the design, but also for installation and operation of the systems which included delivered cost of chemicals. Subsequent implementation has greatly reduced safety and environmental risks with the receiving, storage, and handling of the feedstock, specifically:

- Direct release of WXYZ during unloading, from storage tanks and vent systems to the HCl scrubber (previously producing localized high atmospheric HCl concentrations);
- Reduced incidence of steam explosions from uncontrolled mixing of water and WXYZ (and subsequent generation/release of HCl);
- Better control of scrubber operations (moderated temperatures, lower HCl concentrations, etc.).

## 2.2. The Perils of Not Following the MERITT Approach

The difficulties that arise when all the different paradigms are not involved or coordinated can be even more obvious and instructional than the outcomes of the successful application of MERITT. There are many examples of focusing on IS only to create a waste management issue, or of striving for pollution prevention and creating safety issues. This has created a false sense of conflict or competition between IS and P2, as well as among their practitioners. It is not an inherent conflict,

## Changing States of Knowledge

The initial use of MTBE was heavily driven by regulatory pressures to use this (or another) additive in order to lower emissions. At the time of the introduction of MTBE into most gasolines, it was one of the cheapest and most readily available additives; moreover, the extreme difficulty in removing it from soil and groundwater was not known. However, spills of gasoline containing MTBE from leaking underground storage tanks, among other sources, and releases of unburned fuel from two-stroke engines common in personal watercraft have led to significant soil and groundwater contamination. Now the concerns about MTBE-related contamination are sufficient to drive state and federal agencies to apply pressure to eliminate the use of MTBE.

## Consumer Product Development

A consumer product company was trying to improve and enhance one of its most popular personal care products. At the same time that they sought to increase the market appeal of the product by making it more convenient to use, they decided also to make the manufacturing process more environmentally benign. Unfortunately, the expertise they initially brought to the process redesign effort was not broad enough, and the process that was developed was shown to introduce a health risk to the consumer due to the addition of a new substance in the manufacturing process. The process development cycle was almost complete at the time of this discovery, and major delays and budget increases resulted from having to go back and make a change late in the product development cycle.

but rather a conflict in practice due to the lack of sufficient communication and cross-fertilization. If these two paradigms are applied jointly (ideally as part of a formal approach), solutions can be found that optimize both IS and P2. Similarly, the other paradigms of interest can have their unique perspectives added to the mix to enhance the overall returns.

## 2.3. The Business Case for MERITT

The early incorporation of MERITT in the process development cycle can result in simpler process and equipment design, reduced capital investments, increased reliability, improved equipment effectiveness, reduced long-term staffing needs, increased time efficiencies, higher yields or

3M has achieved $827 million in first-year savings for pollution prevention projects, attributed in large part to alternative approaches to their historic solvent use. Some of these alternatives have offered not only environmental improvements but also pathways to new products.

—*Fortune: Industrial Management & Technology, vol. 142, no. 3, July 24, 2000*

product quality, and, ultimately, higher ROI. Front-end loading within the process development process increases the number of these benefits available and makes them available sooner. Also, MERITT solves issues up-front rather than passing them on to operations, and provides true solutions as opposed to less than optimal fixes. Without MERITT many of the benefits will never be realized, given day-to-day business pressures and a single focus on optimizing for IS *or* P2 *or* GC, etc.

The actual business case for MERITT, as opposed to the individual paradigms, must be largely qualitative at this time, as there are few examples of truly integrated approaches and their application. However, if the long-term issue of making the business case is not recognized and supported from the outset of the introduction of MERITT, it can be very difficult to get the needed and desired comparison performance data. Given the individual benefits already cited, as well as those made in the PSM business case recently put together by CCPS (see Appendix C), it is easy to envision the magnitude of the benefits available—so long as the implementation of MERITT is efficient and does not use resources unnecessarily or inefficiently.

The business case will ultimately need to address the advantages of reduced cycle time, reduced capital and operating costs, increased yields, better product performance/characteristics, etc., as traded off with potential increased capital investments and human resource needs at various stages of the process development cycle. As an additional consideration, concurrent with implementing MERITT, there may be a need to modify any existing IS, P2, GC, DfE, or GT programs within a company to minimize redundancy and overlap in both efforts and resources.

Some of the areas where costs may be eliminated or greatly reduced if MERITT is applied from the beginning (Stage 1) of the process development cycle include:

- Handling hazardous materials and working under hazardous process conditions (training, monitoring, recordkeeping, and PPE may be affected);
- Cleanup costs at decommissioning;

- Regulatory oversight and involvement—a distracting, time consuming, and costly interaction;
- Lost time due to various sources of delay.

If MERITT can lead to increased productivity, it can also help companies deal with existing expansion constraints.

The Michigan Source Reduction Initiative, in which Dow played a leading role, reduced targeted emissions by 43% and targeted wastes by 37%. Some of the projects had one year ROIs as high as 100–1000%. In many cases these rapid returns were a result of reducing not only the wastes associated with certain materials, but also the operating costs of storing, handling, and managing the original material.

—*Fortune: Industrial Management & Technology, vol. 142, no. 3, July 24, 2000.*

## 2.3.1. Sustainability

Sustainable development was defined in 1987 by the Brutland Commission (The World Commission on Environment and Development) as:

Development that meets the needs of the present without compromising the ability of future generations to meet their own needs.

Sustainability concerns will require good, usable metrics (see Section 2.4 and Chapter 7) and will continue to drive new regulatory requirements. They may also be a driver of more integrated tools and approaches, both of which are inherent features of MERITT. This is an area in which AIChE/CWRT has been and continues to be active, through workshops and conferences (see also www.aiche.org/cwrt/projects/).

Sustainability is also a good way to tie in management commitment and may offer a communication pathway for MERITT as a means of actively addressing sustainability concerns.

The true sustainability of a new process must be measured carefully. Growing green plastics directly from corn has been tried by several companies in recent years, but the energy requirements to extract the plastic are much larger than generally realized and currently keep this from being an attractive, let alone sustainable, process.

### 2.3.2. Meeting New Regulatory Demands

There appears to be a virtually continuous addition of new pressures and constraints on the regulatory side, with no letup seen in the near future. Early involvement of the appropriate staff with knowledge of existing and reasonably foreseeable regulatory requirements can help assure that the process is as flexible in accommodating and as responsive to regulatory changes as possible. Adherence to the basic tenets of MERITT can help ensure minimum downstream disruption of the process when changes are enacted.

## 2.4. The Need for Cost and Benefit Metrics to Demonstrate Value

There is a need to start the development and population of metrics now if the business case is to become more quantitative in the near future. For instance, it may be very helpful to collect some data for existing processes to allow comparisons between processes designed with MERITT and those designed without it. Basic or elemental measures can be as or more important than aggregated metrics, and are often more readily quantified.

Establishing the proper metrics is not sufficient; they must be tied to accountabilities and responsibilities or the necessary changes will not be made—whether to MERITT or the process design. Sources for metrics are extensive and more varied than might actually be desired! Table 2-2 lists just a few of the sources to consider.

Discussions follow for two of these efforts that are trying to consider both selected EHS issues and business performance.

### 2.4.1. WBCSD

The World Business Council on Sustainable Development (WBCSD) has proposed both generally applicable and business specific indicators to contribute to a measure of eco-efficiency [*Measuring Eco-Efficiency: A Guide to Reporting Company Performance,* June 2000—available at www.wbcsd.com]. These indicators are further categorized as being related to product or service value or to environmental influence. The five generally applicable indicators for environmental influence all

**TABLE 2-2. Sources of Potential Metrics**

| Source | Types of Metrics |
| --- | --- |
| International Standard on Environmental Performance Evaluation (ISO 14031) | Basic framework for selecting environmentally related measures |
| Guidelines for Integrating Process Safety Management, Environment, Safety, Health, and Quality (1997) | A wide range of potential measures for IS and P2, both leading and lagging |
| Coalition for Environmentally Responsible Economies (CERES) | The Global Reporting Initiative (GRI) focuses on measures that can be used for reporting on corporate sustainability |
| CCPS efforts to develop PSM measures | Detailed process safety metrics |
| World Business Council for Sustainable Development (2000) | Eco-efficiency measures in a report available on the group's web site; more information in text below |
| Organization Resource Counselors (2000) | A task force is looking at alternative metrics for health and safety measures; more information in text below |
| CWRT (1999) | Total Cost Assessment (TCA) methodology manual considers costs associated with E, H, and S issues to develop an overall cost estimate in different categories |

relate to product/service creation (rather than use, which is viewed as business or industry specific):

- Energy consumption
- Materials consumption
- Water consumption
- Greenhouse gas emissions
- Ozone-depleting substance emissions

Although the WBCSD reviewed many potential criteria, they found that those that address energy and material use are the most reliable and robust indicators. Specifically, three of the criteria relate to consumption: material, energy, and freshwater use. Two criteria focus on the associated releases of toxins and general pollutants. Specific definitions have been developed for each of these criteria. Several future criteria have also been suggested, but agreed-upon measurement methods are needed.

The overall set of criteria recognize the importance of the financial component, allow for growth, take advantage of established lists (for toxic and pollutant releases) and are consistent with past programs. The WBCSD believes that the selected criteria are simple, reproducible, and easy to understand, communicate, and use.

## 2.4.2. ORC

Organization Resource Counselors (ORC) has established a task force on alternative metrics for showing the value of EHS programs that can also be a source of good ideas for measuring and monitoring the effects of following MERITT. The focus of this task force has been on metrics that measure process variables and outcomes, to motivate better behaviors and be consistent with other business metrics. Some so-called *trailing* or *lagging* metrics focus on the traditional OSHA incidence rates, but these are balanced with *leading* and financial metrics.

Many different real-time or leading-indicator type metrics are available, both traditional and nontraditional in focus. These need to be sorted through to hone in on a representative and meaningful few that are relevant to a given company or industry.

The financial metrics are also very traditional, but suffer from limited application to EHS costs and benefits and difficulties in quantifying benefits and indirect costs.

## 2.4.3. Recommendations Regarding Metrics

Two categories of metrics are essential, and are discussed in more detail in Chapter 7. These are:

- *Effect-related metrics:* metrics that reflect the benefits of applying MERITT in terms of cost-effectively producing better EHS performance in the developed process. These may include both traditional E, H, or S measures and integrated measures unique to MERITT or based on some of the indices in use or under development. Both leading (or predictive) metrics that can be used at stage gates and lagging metrics that can be used to calibrate with actual project performance will be needed.

- *Process-oriented metrics:* metrics that track the costs of the MERITT approach and support its continuous improvement. These metrics will focus on the utilization rate and costs of integrating EHS resources into the process development process, the costs of developing and applying tools, any influence on project schedule (perhaps by stage), and others yet to be identified. These will generally tend to be lagging measures.

Successful metrics will tie into existing business processes as much as possible, and should also take advantage of and link to AIChE's Total Cost Assessment methodology and other new tools that a company may be using. Chapter 7 identifies a number of specific metrics for consideration.

# 3

# Life-Cycle Stages

The MERITT approach is intended to be integrated into established internal work processes and management systems for process development and modifications to existing operations. This chapter provides an overview of typical process development life-cycle stages and associated work processes that most enterprises employ. By comparing company-specific life-cycle stages with the life-cycle stage definitions provided, an enterprise can interpret how MERITT would be applied to its specific set of procedures and management systems. Sections 3.1 and 3.2 provide an overview and are intended for readers who already work within a formalized process development process culture. Readers who are less familiar with these business processes are encouraged to read the entire chapter.

## 3.1. Introduction

Product or process development or modification begins with a defined need, for example a customer wants improved performance, manufacturing needs improved efficiency, the business identifies a new product, or a regulation/law dictates a change. Most enterprises put business processes in place for implementing, managing, and tracking the development process from inception to operation to deactivation (decommissioning). Enterprises that engage in such activities generally do this well, especially in regard to addressing time to market, capital conservation, life-cycle costs, and to some extent EHS risk management. However, it is likely that EHS risk management is sub-optimal (i.e., the total life-cycle cost is higher than necessary because the best integrated solution for EHS issues was not considered) in many instances due to lack of an integrated EHS approach. The MERITT approach is intended to address this weakness by working within the existing business process structures as applied to process development.

All significant industrial process development endeavors involve a sequence of life-cycle stages (e.g., conceptualization, research, development, piloting/validation, engineering, and manufacturing). In general, most companies and organizations have existing work processes (stage reviews, milestones, project execution, etc.) to manage and control the process development throughout the various life-cycle stages. To facilitate the application of MERITT and ensure its institutionalization, MERITT needs to closely link to these existing internal work processes rather than attempting to institute a completely new approach. This chapter lays the foundation for MERITT by discussing process development in the context of life-cycle stages and work processes.

## 3.2. Phases of Development

The development process is comprised of four phases: research & development, project implementation, production (manufacturing), and postproduction (decommissioning). The development process initiates in the R&D phase, usually as a result of exploratory research (discovery/conceptualization) or a customer need. During the R&D phase, various chemistries and materials are evaluated, process routes are nominated and process steps are defined. In addition, the process is often operated at various scales (i.e., lab and pilot).

Assuming that a process is determined to be commercially and environmentally feasible, the development process will move on to the project implementation phase. During this phase, basic process engineering is performed and reliability/operability issues are resolved, leading to detailed engineering design of a full-scale manufacturing plant. Equipment procurement, plant construction, and pre-startup activities are also accomplished in this phase.

Once the plant is operational, the production phase involves the manufacturing site organization, which has the responsibility for operating, maintaining, modifying, and optimizing the process facility.

The postproduction phase occurs when the plant has achieved its useful life, and the facility is shut down, decontaminated, and mothballed, disassembled, dispositioned, or released for reuse.

## 3.3. Staging and Control

To allow management to exercise technical and financial control, the development process is typically comprised of a sequence (often in series, sometimes overlapping) of life-cycle stages. Control of the development process is accomplished through the use of progress reviews and established criteria and metrics (typically financial and product performance parameters). In many enterprises, the progress review is used as the gate pass to the next stage, and is therefore referred to as the *stage gate* approach [1].

There is no one fixed set of life-cycle stages (LCS) in common use; rather, there is variation among companies. However, to provide a framework for the users of the MERITT approach, a single set of life-cycle stages is presented in this chapter. Throughout the rest of the book, these stages will be used as the structural skeleton upon which the MERITT approach is hung and as the context for application examples.

The stage gate is a mechanism to arrive quickly at a decision about the suitability of moving forward that evaluates whether all criteria have been met to enable starting the next stage. The decision whether to proceed often depends on the state of the knowledge about issues such as:

- Can we really make the product?
- Does it have potential to meet business objectives?
- Can it meet toxicology or regulatory requirements?

The ultimate purpose of stage gates is to avoid committing considerable resources, not the least of which is monetary, to a flawed process, that is, one that will not succeed.

In addition to monitoring the feasibility and progress of the development project, the stage gate process allows for formal handoffs and transfer of accountability at major milestones, such as the transition from the R&D phase to the project execution phase.

For MERITT to be embraced, it should conform to and support existing process development control mechanisms. MERITT should not be implemented in such a way that control of the development process is driven by the approach. Rather, MERITT needs to be integrated in order to enhance and improve the design decisions regarding green chemistry, inherent safety, and pollution prevention. If done properly, it will benefit the development team by eliminating potential show stoppers earlier in the game, and by allowing communication of later stage EHS needs to earlier stage participants. This avoids nonoptimal designs or costly delays due to rethinking design choices late in the project. MERITT can also provide guidance toward better, preferred, and/or more comprehensive options.

## 3.3.1. Generic Stages

The generic development process phase and stages considered in this book are shown in Table 3-1. Enterprises such as those engaged in the development of pharmaceutical and agricultural products also conduct basic research on promising molecules during a "discovery stage." The MERITT approach probably would not be applied during this stage (discovery), but for completeness it is included in Table 3-1. A potential major benefit would be awareness if MERITT were applied at this stage (e.g., facilitating informed choices about raw materials or chemistries).

Some of the major benefits of adopting MERITT are believed to result from application of the methodology in the earlier stages of the development process. Hence, there is a definite front-end loading asso-

*TABLE 3-1. Life-Cycle Process Stages*

| Phase | | Project/Life-Cycle Stage |
|---|---|---|
| Research and Development | 0 | Discovery |
| | 1 | **Concept Initiation** |
| | 2 | **Process Chemistry** |
| | 3a<br>3b | **Process Development or Definition<br>(replication)** |
| Project Implementation | 4 | **Basic Process Engineering** |
| | 5 | Detailed Engineering/Design |
| | 6 | Construction & Commissioning |
| Production | 7 | **Operations (includes upgrades)** |
| Postproduction | 8 | Shutdown, Decommissioning, Disassembly |

Once a plant is running, the temptation is to leave the process alone. In a DuPont process, the plant had been operating for 20 years; however, after a particularly difficult production campaign, the distillation column operation was reviewed with the expectation that the column internals would need to be replaced. On the contrary, the review indicated that the column feed location was incorrect. The conventional wisdom is to locate the feed stream at the tray on which the mixture composition matches that of the feed stream. The better method is to locate the feed on the tray that results in minimum energy consumption; this results in a smaller capital investment and lower operating cost. Relocating the feed to the preferred tray reduced the loss of product to waste from 30 lb/hr to 1 lb/hr, increased column capacity by 20%, and decreased the refrigeration cooling load by 10%.

**The net benefit was a greater than $9,000,000/year increase in revenue to the business.**

ciated with MERITT, meaning that there is considerable focus on using MERITT in Stages 1 through 4 (shown in bold type in Table 3-1). Applying MERITT in stages 6 and 8 (shaded) is least likely to have a major impact. Some impact is also possible in Stage 5, particularly for inherent safety concepts.

Stage 7 presents an interesting opportunity for MERITT associated with process improvement upgrades. Typically, upgrades tend to be add-ons or reconfigurations within the context of the basic process scheme. However, by allowing thinking outside the box (e.g., changing catalyst system, modifying process chemistry, adjusting equipment selection) improvements with very significant EHS and commercial benefits can be achieved. A further discussion of the use of MERITT in this context is provided in Section 3.7.

To allow interpretation of the stages presented, the main characteristics are shown in Table 3-2.

Stage 3 has been differentiated into:

- *3a—Process Development (new process or process redesign).* Stage 3a is associated with internal process development projects and involves pilot plant scale operation for process optimization and generation of engineering design information.
- *3b—Process Definition (plant replication).* Stage 3b is related to new or major capacity addition (replication) projects utilizing

**TABLE 3-2. Traditional Stage Attributes**

| Project/Life-Cycle Stage | | Scale of Operations | Stage Focus | Traditional Process-Related Disciplines* |
|---|---|---|---|---|
| I | Concept Initiation | Bench-top | Product Synthesis | CR, H |
| 2 | Process Chemistry | Lab Scale-up | Basic Chemistries | CR, CP |
| 3a | Process Development | Pilot | Process Optimization | CP, PDE, S, M |
| 3b | Process Definition (replication) | Commercial | Technology Selection | PDE, PE, CPE, S, E, M, CP |
| 3c | Process Definition (upgrades) (originates in L-C Stage 7) | Commercial | Process Improvement | PM, PE, DE, E, S, M |
| 4 | Basic Process Engineering | Commercial | Process Design | PM, PE, DE, S, H, E, M |

* Before MERITT

**Discipline Key:**

| | |
|---|---|
| CR | Chemist—Research/Synthesis |
| CP | Chemist—Process |
| CPE | Conceptual Process Engineer/Economic Evaluator |
| DE | Design Engineer |
| E | Environmental Engineer/Advocate |
| H | Health Specialist |
| M | Material Specialist |
| PE | Process Engineer |
| PDE | Process Development Engineer |
| PM | Project Management |
| S | Process Safety Specialist |

acquired process technology from a process licensor, toller, or joint venture partner. Because licensing agreements may be involved, there could be conflicts with MERITT driven concepts related to process modifications.

- *3c–Process Definition (plant upgrades).* Stage 3c considers the process improvement or upgrading project that is initiated during life-cycle Stage 7.

Process definition, Stages 3b and 3c, is usually the initiating stage for the project and the desired entry point for applying MERITT.

Table 3-2 also indicates the technical disciplines generally involved in the process development at a particular stage. One of the objectives of integrated EHS (or MERITT) is to get the correct disciplines interacting at the most appropriate time. This does not happen often enough, either because appropriate disciplines are excluded from certain stages, or because of compartmentalism within the stage. The MERITT approach is intended to challenge the way that process development teams currently interact. Associated with each of these stages are corresponding process characteristics expressed in terms of objective, scope, approach, and deliverables in Table 3-3.

## 3.3.2. Resource Allocation and Control

To track progress, allocate resources, and control costs, organizations create management systems for process development. These invoke a variety of mechanisms including assigning development teams, empowering leaders and practitioners, and establishing periodic management reviews (stage gates). There are different approaches for achieving acceptable results.

One major enterprise surveyed uses seasoned development leaders who are given broad management authority to drive the process. It is the development leader's responsibility to ensure that EHS issues are addressed during his or her phase of the project. Other chemical companies prefer to invoke procedures and training that institutionalize the use of development project methods that address EHS issues.

An important mechanism in the process development management process is stage gate control (irrespective of MERITT). Stage gates are inserted into the process to challenge the development team and ensure that the initial objectives regarding product efficacy, quality, process efficiency, EHS considerations, financial return, and cost are still achievable. Figure 3-1 illustrates the overlay of the stage gate process on the time line of the development stages. The development stages are shown overlapping, as this is typical when the process development cycle is compressed. Therefore, the stage gate review will often take place before the current stage is completed, in order to allow the next stage to begin. This can present issues for the development process and the inclusion of MERITT, which are explored in Section 3.6 of this chapter.

At each stage gate, the project has to demonstrate the ability to meet certain objectives and criteria. For the most part, stage gate metrics are

**TABLE 3-3. Description of the Process Characteristics for Each Project Stage**

| Project Stage | | Process Characteristics |
|---|---|---|
| **Stage 1** **Concept Initiation** | Objective | Refine product and synthesis requirements |
| | Scope | Laboratory testing (bench-top scale batch) |
| | Approach | Product isolation and properties testing and screen chemistries |
| | Deliverable(s) | Product formulation and synthesis route(s); conceptual economics |
| **Stage 2** **Process Chemistry** | Objective | Select basic process scheme |
| | Scope | Laboratory testing (lab scale-up batch and limited multiunit continuous) |
| | Approach | Basic kinetics studies followed by validation testing of core process steps |
| | Deliverable(s) | Basic process flow schematic; preliminary operating units and conditions |
| **Stage 3a*** **Process Development** | Objective | Finalize unit operations and optimize process design and operating parameters |
| | Scope | Pilot scale testing supported by selected laboratory studies |
| | Approach | Variational testing of subsystems followed by demonstration runs of integrated equipment |
| | Deliverable(s) | Concept design—preliminary PFD and M&EBs; equipment capacities and materials |
| **Stage 3b*** **Process Definition (Replication)** | Objective | Select process technology and optimize process design and operating parameters |
| | Scope | Process simulations supported by plant operating experience |
| | Approach | Evaluation of alternative process conditions, raw material compositions, throughputs |
| | Deliverable(s) | PFD and M&EBs; equipment capacities and materials |
| **Stage 3c*** **Process Definition (Upgrade/Mods)** | Objective | Obtaining business value from process improvement |
| | Scope | Improving process efficiencies and facility availability, eliminating waste (all types) |
| | Approach | Evaluation of process materials and utilities utilization, and operating reliability |
| | Deliverable(s) | PFDs; M&EBs; P&IDs; equipment selection; general arrangements; control schemes |
| **Stage 4** **Basic Process Engineering** | Objective | Finalize process design in support of ±20% cost estimate |
| | Scope | Complete 35% level engineering/design |
| | Approach | Evaluation of alternative equipment and configurations/layouts/constructability; vendor quotations |
| | Deliverable(s) | PFDs; M&EBs; P&IDs; equipment definition; general arrangements; control schemes |

*Only 3a, b, or c applies to a given project

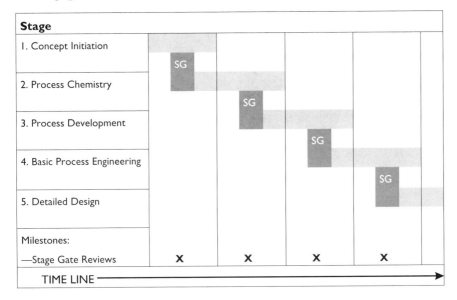

*Figure 3-1. Process development stages and reviews.*

related to schedule (time to market), financial (process economics), regulatory issues (e.g., FDA approval) and whether product performance levels are being achieved. Existing EHS metrics tend to be nonintegrated metrics (addressing E, H, or S) and focused on outcomes or results. Thus, they are seldom used as stage gate criteria. The few EHS criteria in use are more pass/fail oriented (Does the process avoid using chlorine?), and do not serve to enhance the project or motivate changes. Hence, the stage gate process is both a challenge and an opportunity for getting meaningful integration of EHS benefits into the process development management processes. Section 3.4 provides some guidance for meeting this challenge.

### 3.3.3. Interpretation of Stages

While the stages and processes described throughout this book are unlikely to identically match your company's management systems, they should provide some guidance for mapping the activities involved onto your existing internal work processes. During this mapping, you should recognize that:

- Some stages may be further combined or disaggregated, especially during the initial phase of development;
- The stage gates may occur at different points in the overall time line; and
- The overlap of stages may occur more or less routinely in your organization (see Section 3.6 on fast track development).

A comparison of the stages outlined in this book with those used by others is shown in Table 3-4 to illustrate how they align. The second column in the table shows the stages outlined by the *IN*herent *S*HE *I*n *De*sign (INSIDE) Project, a joint industry–European Union Industry Safety research project that began in 1994 (details in Chapter 5), and the third column contains stages as defined by a major petrochemical company. The last column of the table is left blank to allow readers to map their own system of stages. Note that there is considerable consistency among the stage frameworks shown, except that the INSIDE Project did not subdivide plant design into basic and detailed design.

**TABLE 3-4. Stage Alignment**

| Generic | INSIDE[1] | Company "A" | Your Company |
|---|---|---|---|
| I<br>Concept Initiation | I<br>Preliminary<br>Chemistry Routes | I<br>Preliminary<br>Assessment | |
| 2<br>Process Chemistry | 2<br>Chemistry Route<br>Detailed Evaluation | 2<br>Detailed Assessment | |
| 3<br>Process Definition or<br>Development | 3<br>Process Optimization | 3<br>Development | |
| 4<br>Basic Process<br>Engineering | 4<br>Process Plant Design | 4<br>Basic Design and<br>Appropriation | |
| 5<br>Detailed Engineering | 5<br>Process Plant Design | 5<br>Detailed Engineering<br>and Procurement | |

[1]INSIDE Project is a joint industry–European Union Industry Safety research project that began in 1994

Naturally, some projects (e.g., plant upgrades) will not progress through all of these stages. The MERITT approach described in Chapter 4 is designed to be adaptable to a wide variety of situations and process/product development stage systems.

## 3.4. EHS Constraints and Opportunities

### 3.4.1. Stage Constraints

As projects progress through the development stages, process materials, operating conditions, equipment requirements, and control parameters become more defined, such that by Stage 4 the chemistry route, raw materials, process conditions, and process steps have been essentially fixed. By then, the ability of any integrated EHS methodology to make a significant difference is severely limited. Or, as illustrated by the examples in Chapter 2, application of EHS this late in the development process can prove to be very embarrassing to the development team, and in some instances, can result in termination of the project. Therefore to derive the most benefit from MERITT, the concepts it embraces need to be incorporated at the earliest stage of serious development work. This is demonstrated in Figure 3-2.

> In the pharmaceutical industry, the point of fixing all process conditions occurs early in the R&D phase, so that clinical trials needed for FDA approval can be started to meet the overall development cycle schedule goal. Other industrial sectors like cosmetics and paint also face similar limited timeframes for improvements.

Constraints can result from decisions in earlier stages that limit the choices available for addressing EHS concerns. For example, a decision by a process chemist to employ a chlorinated solvent severely limits the environmental/process engineer's choice of disposal method for waste solvents. Similarly, the number of process steps using different solvents and/or process conditions can significantly affect process energy consumption and water usage rates, both of which have environmental and possibly inherent safety impacts. At the earliest development stage, it is not the intent of MERITT to impede progress; however, the development practitioners need to be sufficiently informed to make conscious choices

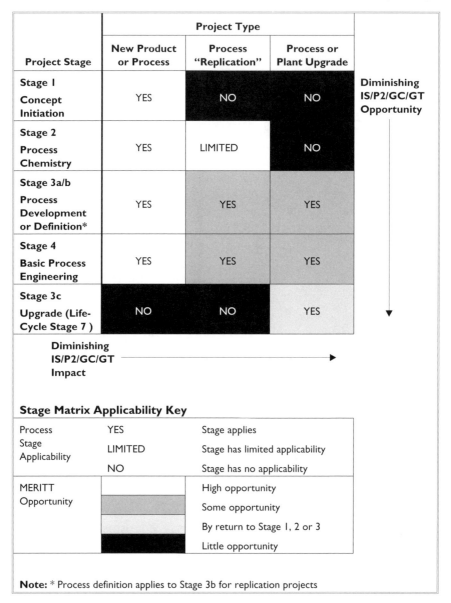

| Project Stage | Project Type | | | |
|---|---|---|---|---|
| | New Product or Process | Process "Replication" | Process or Plant Upgrade | |
| Stage 1 <br> Concept Initiation | YES | NO | NO | Diminishing IS/P2/GC/GT Opportunity |
| Stage 2 <br> Process Chemistry | YES | LIMITED | NO | |
| Stage 3a/b <br> Process Development or Definition* | YES | YES | YES | |
| Stage 4 <br> Basic Process Engineering | YES | YES | YES | |
| Stage 3c <br> Upgrade (Life-Cycle Stage 7 ) | NO | NO | YES | |

Diminishing IS/P2/GC/GT Impact

**Stage Matrix Applicability Key**

| Process Stage Applicability | YES | Stage applies |
|---|---|---|
| | LIMITED | Stage has limited applicability |
| | NO | Stage has no applicability |
| MERITT Opportunity | | High opportunity |
| | | Some opportunity |
| | | By return to Stage 1, 2 or 3 |
| | | Little opportunity |

**Note:** * Process definition applies to Stage 3b for replication projects

*Figure 3-2. MERITT applicability and opportunity by project stage.*

and to understand the implications of their choices. An effective way to influence the application of MERITT is by engaging empowered leaders and practitioners (including gatekeepers at stage gates) in adhering to its tenets and achieving its benefits and following through consistently on all phases of commercialization.

Another constraint is a general lack of awareness of integrated EHS concepts and their value. Quite often the insertion of various EHS tenets into some stage of the development process is a result of having a strong internal proponent. Relying on such an approach is problematic if the champion moves on to other activities. One of the aims of the approach presented in this book is to encourage the adoption of MERITT in a more universal and systematic way early in the process; not merely relying on one individual to drive it (*pull* is much better than *push*).

Stage gate metrics that reflect EHS values are very desirable, and the lack of these criteria and integrated metrics is another potential constraint. The stage gate process and criteria influence how the development team approaches its project. In one company, the project is assigned a "review coordinator," whose responsibility it is to ensure that all stage gate requirements are adequately addressed. In most companies, the EHS expectations for proceeding to the next stage are not explicit. Hence, the development focus will gravitate toward the explicit criteria for product performance, schedule attainment, and financial aspects.

More explicit EHS criteria or metrics that compare the inherent safeness and environmental sustainability of alternatives should be considered for use in stage gate reviews. This will help institutionalize MERITT, because the development team will anticipate this as a requirement for the project to proceed.

## 3.4.2. EHS Opportunities

EHS opportunities are inexorably linked to the development stage and the corresponding process characteristics. Based on the process characteristics shown in Table 3-3, a mapping of IS, P2, and GC opportunities by project stage is provided in Figure 3-3. In general, the opportunity for IS/P2/GC concepts is greater at earlier project stages. Because GC recognized the need to minimize the EHS effects of chemical processes by addressing the fundamental chemistry employed, it is heavily weighted

| Project Stage | GC | IS | P2 |
|---|---|---|---|
| Concept initiation | | | |
| Process chemistry | | | |
| Process definition or development | | | |
| Basic process engineering | | | |
| Detailed engineering | | | |

Selective opportunities

Substantial opportunities

Major opportunities

*Figure 3-3. EHS opportunities.*

toward the early stages of process development. Similarly, IS emphasizes changing the nature of the process in the early stages in order to
reduce or eliminate the hazards associated with the materials and operations in such a way that the reduction is permanent and inseparable.
However, to be effective, IS is best applied when some process definition
is available. P2 generally addresses upstream waste reduction and has
found countless applications in manufacturing, process, and service
industries. P2 has a heavy environmental focus that in the process
industries is most closely aligned with environmental and chemical
engineers, and therefore has most frequently been used in mid- and
later-stage development efforts and upgrade projects.

Representative opportunities associated with various process
aspects are shown in Table 3-5 (Figure 1-3 defines many of the terms
used in the table). A more detailed list of opportunities by stages is provided in Chapter 4.

## 3.5. EHS Information Needs

### 3.5.1. Requirements

The type, quality and extent of information available at each stage can
have a major impact on the success of implementing an integrated EHS
approach. As shown earlier in Table 3-3, the level of process definition

*TABLE 3-5. Representative EHS Opportunities*

| Process Aspect | Opportunities |
| --- | --- |
| Materials/Resources | Alternate reagents |
| | Alternate solvents |
| | Alternate catalysts |
| | Alternate raw materials |
| | Waste/byproduct reuse |
| | Recover/recycle solvents |
| | Raw materials modification |
| | Recycle raw materials |
| Conditions | Alternate catalyst system |
| | Reaction heat sink (Moderation) |
| | Moderate conditions (P, T, pH) |
| | Adjust concentrations (Moderation) |
| | Transform waste (Alternative waste treatment) |
| Equipment/Containment | Combine steps (Minimization/Simplification) |
| | Fewer reaction steps |
| | Total containment design |
| | Reduce equipment size (Intensification) |
| | Improve constructability |
| | Continuous versus batch operation |

and therefore the amount of information available steadily increases with each successive stage. In the earliest development stage (Stage1), it is expected that an integrated EHS approach would rely heavily on the information typically developed at that point. However, MERITT should attempt to influence the development of information needed for integrated EHS in later stages (2 and 3) of development. This will help eliminate ineffectual interactions among the development team and the EHS advocates due to lack of sufficient information. A more detailed discussion of MERITT information needs is provided in Chapter 4.

## 3.5.2. Anticipating Information Needs

Information requirements must be anticipated well in advance of the stage in which they will be needed, otherwise the progress of a project can be seriously and negatively impacted. A benefit of applying integrated EHS at the earliest stage of development with the appropriate participants is early communication of EHS data needs for subsequent stages. The anticipation of data needs should begin at Stage 0 and should be continued through all subsequent stages.

> In one instance, during the process development stage a development team failed to anticipate the data needs for the basic process engineering stage. This resulted in the need to repeat pilot plant runs to obtain the data, causing a significant delay in the project execution phase.

## 3.6. Fast Track Development

Along with the globalization of business has come an urgency to get products to market ever faster. Development cycle time has been reduced and many projects are put on the fast track. The impact of fast tracking on the development process and project management is schedule and stage activity compression. The result of schedule compression is stage overlap, with the decision to start the next stage occuring before the present stage is completed. Stage activity compression is focused on reducing the time period for completing a stage by performing activities in parallel to the extent possible.

> There is a widespread (but generally erroneous) belief that the more urgent the situation, the more you can ignore inherent safety, pollution prevention, etc. in developing the solution. This creates a desire or perceived need to have a bypass to regular requirements. MERITT will help streamline the process and limit such problems and attitudes.

The application of an integrated EHS approach needs to anticipate issues associated with fast tracking. An obvious challenge caused by schedule compression is that the time interval between key decision points is significantly shorter. This translates into less time to identify, evaluate, and validate EHS options. Activity compression (parallel processing) can impact data availability, in that not much data may be available until the end of the stage. When it becomes available, there is a rush to assimilate the results and move to the next stage. Another potential issue associated with activity compression is that many resources may be busy on other tasks, but this can be alleviated through a focused prioritization of workloads. Under such time pressures, there will likely be resistance to spending considerable additional effort or time to apply approaches such as MERITT, especially if the approach involves assembling sizable teams.

Therefore, implementation of an integrated EHS approach should address ways to overcome or reduce the constraints of fast track development. This will entail the incorporation of EHS thinking and concepts into easily applied tools such as guidelines and "think lists" to be used by development practitioners. The conditions that are required for handoffs from one stage champion to the next (if different champions are responsible for each stage) also needs to be considered. The requirement for continuity and ease of communication provides an opportunity for incorporation of an integrated EHS philosophy during procedure or tool development, and allows substantial implementation by the individual chemist or development engineer. Furthermore, forced prioritization of workloads can be used to offset activity compression in fast track situations.

## 3.7. Plant Upgrades and Modifications

Plant upgrades and modifications can present a unique opportunity for incorporating integrated EHS concepts, provided the project is allowed some latitude to broadly consider different process alternatives, such as catalyst improvements, different reactor systems, substitute solvents, etc. The example in Chapter 2 of an old chemical manufacturing facility scheduled for relocation demonstrates how looking at the issues with an integrated approach can be very beneficial.

### 3.7.1. Stage Iteration

Some plant upgrade/modification projects originate from R&D, but many are defined as improvement projects from the operations organization (life-cycle Stage 7). Typically the project starts with Stage 3c type

The manufacturers of a herbicide intermediate were unable to meet demand when operating their batch process at full capacity. To overcome the production shortfall, they purchased the intermediate from a competitor at a price higher than their cost of manufacturing. Duplication of the existing batch process seemed the conservative approach to increasing production. However, the business team saw an opportunity to meet the expansion objectives with minimal investment by changing from batch to continuous reaction technology. The team persevered through many intensive technology and project reviews, and concluded that the continuous process appeared inherently safer and technically viable. Within eleven months from concept to startup, the business team achieved a 240% increase in production capacity when compared to the original batch process. In addition, the continuous process demonstrated a 29% reduction in methanol emissions per pound of product when compared to the batch technology.

activities (see Table 3-3) and progresses through life-cycle Stages 4, 5, and 6. The project scope is usually quite specific and somewhat narrowly defined. EHS opportunities may be limited to inherently safer equipment designs and end-of-pipe pollution control and abatement options. Hence, potentially large value-creation opportunities that could be found by using an integrated EHS approach are not identified. However, there is evidence from company success stories that significant opportunities exist to those willing to challenge their conventional thinking.

The way to find such opportunities is to use the opportunity of an upgrade/modification project to look at the situation in the broader context of EHS values and objectives. Applying MERITT will initiate thinking about other alternatives that can accomplish or exceed the initial improvement target and better meet EHS objectives. Some of the options may be more complex and costly (in terms of capital expenditure) than was originally envisioned. However, the resulting value-creation and savings in operational expenses may easily justify the additional expenditure. In some cases, there is both a capital and operating cost savings, as was demonstrated in the Chapter 2 relocation example.

Obtaining full benefits of the MERITT approach for such projects will likely involve iteration back to an earlier life-cycle stage such as 3a, where some pilot plant testing of a concept could occur. In some cases, returning to even earlier stages (2 or possibly 1) may be feasible, if the value creation potential was very significant. Again, the DuPont project described in Chapter 2 illustrates this concept. Switching from a fixed-bed to a fluidized-bed reactor system represented a significant scope change and required iteration (from Stage 3c—Process Definition to Stage 2—Process Chemistry and Stage 3c—Development) involving bench scale and pilot testing. However, the fluidized-bed system reduced compression energy requirements by 85%, in addition to the emission reduction benefit.

Implementing such an approach on upgrade projects will not be easy, due to conflicting priorities of schedule, cost, and staffing. Having a shared vision that it is worth the effort (because benign design is good and can be profitable) is a good place to start. After utilizing the MERITT approach, the consensus may be that other options offer only marginal EHS improvement and are too disruptive to the project schedule and budget. So, at a minimum, the right disciplines and experience will be brought to bear and the decision process will be improved. Alternatively, a significant improvement, which everyone can support, may be demonstrated.

Clearly, the less impact on project schedule and budget that is perceived, the greater the chance that MERITT thinking will be accepted. Fast tracking (discussed in Section 3.6) can be used to reduce schedule impact. During engineering, this may require placing a hold on the design in certain areas (e.g., waiting for completion of Stage 2 or 3) while the rest of the design proceeds. Getting management's commitment for a less proven technology approach may also be a significant hurdle. Demonstration of the concept at lab scale would be appropriate before initiating a major project based on that technology.

## 3.7.2. Creating MERITT Opportunity

Creating MERITT opportunities involves first understanding the origin of the EHS issues, and then challenging the reasoning behind the current process arrangement/configuration. This requires thinking (and perhaps implementing) outside the box and can result in having to go back to earlier development stages. Questions such as "Why are we doing it this way?" or "What if we do it this way instead?", when asked in the proper forum and context, can become the catalyst for significant improvement [2].

Water-soluble solvents from a solution polymerization process were water scrubbed from an air stream. Recovery of the solvents from the water stream was considered to be too expensive, so the water stream (now a RCRA hazardous waste) was incinerated. An extensive review of the vapor–liquid equilibrium data and a pilot plant test showed that the solvents could be separated from the water stream by distillation followed by extraction. The distillation step separated the three solvents with one solvent going overhead into the distillate with the water, and the other two solvents remaining in the bottoms. The solvent in the water stream was then extracted from the water with a low-boiling immiscible hydrocarbon, and the solvent was recovered from the hydrocarbon by an azeotropic distillation column.

**By recovering more than 10 million lb/yr of solvents and reducing the waste incineration load by more than 4 million lb/yr, the new capital investment had only a two-year payback period.**

## 3.8. References

1. Cooper, R G., *Winning at New Products,* Reading, MA: Addison Wesley, 1986.
2. Dyer, J.A., and Mullholland, K.L., "Follow This Path to Pollution Prevention," *CEP*, January 1998.

# 4

# The MERITT Approach

This chapter is written for those individuals who will help with the integration of MERITT into existing process development processes, not for EHS specialists who may serve as the key resources to make MERITT effective in practice. The basic strategy of MERITT in advancing the integration of EHS into the development process derives from a very simple precept—get the right knowledge into the right hands at the right time in the development process. This involves not only driving things earlier into the process but also creating a continuity throughout the process and most importantly, infusing E, H, and S perspectives concurrently. To accomplish this, the overall framework of MERITT is constructed in three parts—*Fundamental Principles,* which form the overarching tenets; *Resource Components,* which comprise the five basic building blocks of MERITT; and *Implementation Elements,* that are five interactive steps that can be used for guiding the process. This chapter discusses each of these and describes how they can be utilized in different types of process development efforts.

## 4.1. The Foundation of MERITT

MERITT derives from a very simple precept—*get the right knowledge into the right hands at the right time in the development process.*

The strategy of MERITT has three parts:

1. Get the broadest range of concurrent EHS perspectives involved in the development process at the earliest possible time.

2. Present EHS perspectives as a unified set of integrated issues rather than independent E, H, and S considerations.

3. Continue this throughout the development process to anticipate needs most effectively and resolve issues along the way.

The survey of industry best practices conducted at the start of this project showed that companies leading the way in integrating EHS practices into process design were achieving significant value from these efforts. Their success spurred interest in further development of these holistic, integrated evaluations. However, this survey uncovered no particular approaches that could easily be adapted for use by others. The early work on this project also identified a need and desire to find a successful way to work within existing processes and to avoid creating new teams or additional time demands.

Success of any methodology, of course, is predicated on the underlying commitment of individual team members involved in implementing the methodology to use pertinent information and act on the results of the evaluations undertaken. Only through early, concurrent, and consistent coordination of EHS considerations can we expect to reduce the uncertainties and avoid the inefficiencies associated with independent E, H, and S evaluations.

To accomplish this, new development processes are not necessarily required, but it is necessary to expand or adjust the processes that already exist. MERITT has not been developed as a new, stand-alone process development methodology, although it can be used to formulate new development processes. It is particularly well suited to processes structured in sequential stages with fairly well-defined stage gates, but it is applicable to almost any development format. The construct of MERITT is more along the lines of an *enabler* than that of a traditional methodology. Because there are many variables in development processes (e.g., the structure of the development process itself, the types of industries and their specific EHS priorities, and the business imperatives associated with any particular project), MERITT has been crafted with the flexibility to ensure that the basic approach can be tailored to accommodate different requirements.

The following sections describe the MERITT approach and provide guidance on successful integration. The discussion is intended to lay out a generic framework that can basically fit into any existing development process. After discussion of the MERITT framework, two cases are examined for the potential application of MERITT. These cases consider the possible range of applications, from a fully orchestrated development process involving multiple stages to a fast-track project involving parallel tracking and quick decision making.

## 4.2. MERITT Approach Overview

### 4.2.1. The Basic MERITT Framework

MERITT is framed in three parts (see Figure 4-1). The first part embodies the *Fundamental Principles* that form the glue that holds everything together. The second part comprises the *Resource Components* which are the five basic building blocks of MERITT, serving as the key assets that support the integration effort. The third part contains the *Implementation Elements* which are a set of five steps used to generate the tactics for utilizing the resource components. These steps help to structure the direction and interrelationship of activities. The resulting tactics will vary depending upon the nature and requirements of a particular development process.

Each of these parts is necessary, and none is sufficient by itself. Successful EHS integration requires all three, however they are manifested and utilized. This is where MERITT's flexibility comes into play, as will be described in more detail.

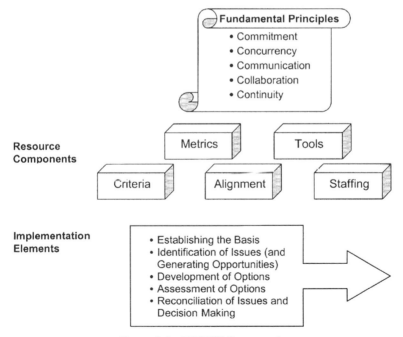

*Figure 4-1. MERITT framework.*

Much of this framework has been derived from two types of sources. First are P2, DfE, GC, and IS program concepts that have already been developed and fairly widely accepted. The second are best practices of leading companies that have been engaged in advancing the integration of E, H, and/or S into corporate policies and procedures. In many companies, much of what is required already exists, especially where programs such as P2, DfE, or GC have been actively undertaken. It is not intended that the existing infrastructure and supporting constituents be rebuilt or replaced. What works well should certainly be sustained and, at most, expanded as appropriate to envelop the broader reach of an integrated EHS perspective. In this regard, there is a great deal of latitude in the nature and content of the resource components and in how tactics are developed. Indeed, these should be crafted to meet the specific needs of each enterprise. It is necessary only that they exist in one workable form or another.

### 4.2.2. The Five "C" Fundamental Principles

The importance of the five fundamental principles cannot be overemphasized. Each is instrumental to the success of EHS integration.

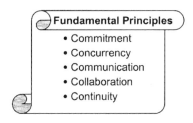

**Fundamental Principles**
- Commitment
- Concurrency
- Communication
- Collaboration
- Continuity

1. *Commitment*—There must be commitment from the process development leaders, the project managers, the business managers, and the corporate leadership. This is ultimately a top down requirement that must support and fit within the corporate culture for positive reinforcement of the principles, rather than internal conflict.

> Clearly articulate business value through success (and failure) stories as well as examples. Ensure that individuals' metrics as well as corporate reward and recognition systems also reinforce desired behaviors.

2. *Concurrency*—Thinking about environmental, health, and safety concerns concurrently while identifying issues, setting priorities, and defining and actualizing opportunities is the cornerstone of the approach. Concurrency is all about *fusion* and *infusion*— fusion of environmental, health, and safety perspectives, and infusion into the development process. Several of the other principles directly bolster concurrency.

> Draw from existing in-house joint EH, ES, HS, and EHS projects and practices, such as in facility auditing and life-cycle assessment work, as examples of how these disciplines can successfully work together. Emphasize the criticality of getting the right input in a timely fashion.

3. *Communication*—Both concurrency and collaboration require communication. Throughout the development process there must be open and ongoing communication to get the right mix of expertise and knowledge involved at each step of the development process; exchange ideas and resolve issues; and disseminate information to all active participants, business leaders, and customers.

> Focus communication efforts on the important issues. Make them valuable and short. Do not inundate people with day-to-day developments and "newsy" narratives.

4. *Collaboration*—Coordination of activities is clearly necessary but not itself sufficient in achieving concurrency. There must be collaboration. Individuals must work together in identifying, addressing, and resolving EHS requirements. It is important to note that collaboration does not require formal meetings. Often smaller working groups are much more successful.

> Collaboration (and buy-in) can be successfully effected through the simple act of joint development and/or use of resource components, such as tools or metrics, discussed in the next section.

5. *Continuity (Concept to Completion)*—Considerations and evaluations must start at the earliest possible stages of the project and the effort must be carried through successive development steps. In essence, a *continuum* for organized thought and information transfer must be created that will lead to informed decision making at each step so that each successive effort can build upon what has already been done. By the same token, up-front work must anticipate the future needs in the development process for the work to proceed efficiently. This necessitates collaboration.

> Avoid "upstream" barriers. Getting "downstream" development people involved earlier in the process can provide preemptive feedback that can help avert information gaps or precluded options later. Similar experience from other projects can also help.

Many leading companies have already embraced most of these principles within corporate environmental and/or safety programs, however, most have yet to achieve complete success in the areas of collaboration and continuity. There are even fewer companies, if any, that fully address the integrated EHS perspective. Progress in these areas will require both dealing with a company's formal institutional structure (e.g., organizational format, management style, and information systems) and contending with the company's culture and customs (e.g., unwritten rules and legacies of entitlements). An effective on-going training program is generally regarded as essential to support changes of management systems, procedures, or behaviors.

## 4.2.3. Resource Components

The following discussion of the five resource components is intended to provide an introduction to the basic concepts. More familiarity with these will be gained from descriptions of applications of MERITT and from the more in-depth information presented in Chapters 5 and 7.

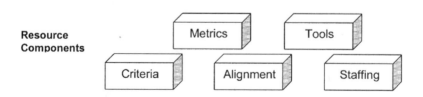

*Criteria*

Two levels of criteria are required. First, there are the criteria or specifications that set the standards for the process design. These define the minimum requirements in all aspects of the project. Make certain that they accurately reflect the goals and objectives established for the project (or each individual stage). They can and should be continually revisited for each project and at the start of each formal development stage.

Examples of typical project criteria include a number of interdependent factors:

- Cost
- Financial performance (e.g., ROI)
- Product quality and performance
- Process operability and maintainability
- Process reliability
- EHS performance
- Schedule
- Adherence to other corporate or business unit imperatives (e.g., design standards)

We are primarily concerned here not only with the optimization criteria specific to EHS issues, but also with the overall optimization criteria and how they might impact EHS-related decisions.

> If a developmental chemist selects a chlorinated solvent to improve reaction yield, this choice may create the need to use more exotic and problematic materials in a downstream thermal oxidizer. Such implications should be considered at the time the solvent is chosen in order to achieve true optimization.

EHS considerations must be merged within the entire development process. They cannot stand above or outside other considerations. It is crucial that all the criteria be clearly defined at the outset and that they be used in both EHS and other project evaluations. There are many sources for identifying potential criteria:

- Corporate/business unit policy
- Corporate/business unit standards

- Regulations (environmental, health, safety, product, etc.)
- Industry standards and codes
- Advocacy acceptance
- Other project specific requirements

> In developing the criteria for any process development program, it is not sufficient to replicate the criteria set from prior projects or simply accept corporate and regulatory prerequisites as a complete set. Criteria setting must encompass the full range of potential factors.

The second type of criteria sets the standards for the development process itself. These relate to how well the process works—e.g., the efficiency in achieving the desired results; how well it truly integrates E, H, and S considerations; and how well it addresses and resolves the predicaments of opposing or contradictory perspectives. This on-going self-evaluation is crucial to winning acceptance and improving the overall development process.

*Metrics*

The criteria alone are of limited value without metrics by which to gauge success. Consequently, there must be two types of metrics that generally align with the types of criteria delineated above. First, there must be metrics for ranking or evaluating the process design at each decision point and to support decision making along the way. Second, there must be metrics that measure the effectiveness of the development process itself. All metrics should provide information that is actually used.

The most important advice when developing metrics is to keep them simple and targeted at their ultimate use. Metrics should serve as a means of determining the degree of balance between costs and benefits, as illustrated below.

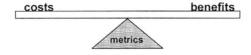

The greatest challenge in developing metrics is that there is still no common currency in the quantification of E, H, and S impacts except cost, and cost is an imperfect measure.

There are some recently developed evaluation tools with embedded, relative ranking systems for several dimensions of inherent environmental, safety, and health concerns, such as the Solvent Selection Guide advanced by GlaxoSmithKline. One potential alternative is to gauge EHS performance in terms of relative ranking systems that relate performance to compliance and corporate threshold limits.

## Tools

There are dozens of tools that have been developed to assist in E, H, and S evaluations. Roughly two dozen have been identified for potential use in supporting MERITT. These have been drawn from several sources, notably:

- Existing E, H, and S paradigms (e.g., P2, GC, IS, and DfE)
- Recent efforts in developing integrated EHS tools (e.g., INSET)
- Codified process safety tools (e.g., PHAs)
- Widely utilized business analysis methodologies (e.g., LCA)
- Decision theory

These tools have been grouped into three general categories for MERITT—inquiry, option generation, and option evaluation/decision support. Inquiry tools are primarily intended to identify EHS issues and deal less with immediate solutions. As the name implies, option generation tools help in the formulation of alternative design concepts to address issues found with inquiry tools. Evaluation or decision support tools focus on methods for ranking alternatives using various EHS and cost criteria. The results are typically expressed in the form of indices or numerical rankings. Figure 4-2 provides examples in each category.

| Inquiry | Option Generation | Evaluation/Decision Support |
|---|---|---|
| Brainstorming | API 752—Siting | Dow/Mond Index |
| Chemical Hazards Analysis | Brainstorming | Fault Tree Analysis |
| Chemical Interaction Matrix | IS Strategies | E, H, or S Indices |
| HAZOP | P2 Strategies | Life-Cycle Analysis/Cost |
| What-If | Reaction Hazard Analysis | Levels of Protection Analysis |
| Dow/Mond Index | Solvent Selection Guide | Process Simulation |
| E, H, or S Indices | | Solvent Selection Guide |
| Reaction Hazard Analysis | | Total Cost Assessment |

*Figure 4-2. Examples of MERITT tools categorization.*

Note that tools may appear in more than one category because they can often be used in different ways or with different emphasis. A detailed discussion of tools is provided in Chapter 5.

### Alignment

Alignment is unlike the other four components in that it is imbedded within all of them through their creation and utilization. It addresses who does or shares what, with whom, at each stage gate and is manifested in several dimensions:

- E, H, and S functions within the organization as a whole
- E, H, and S activities within the project team
- Activities between stages
- The project team with the business unit
- The project team with the corporate organization(s)

The first three of these are the focus of integrating the different EHS disciplines and then taking the "unified EHS" and integrating it into the development process (presuming that the development process is already fairly well established). Unfortunately there is no universal answer as to how best to address these issues. As noted above, tailoring to each organization and to the development process is required. Trial and error in reaching acceptable compromises is inevitable. Nowhere are the five principles of concurrency, collaboration, communication, continuity, and commitment more important to the overall process than here.

> Alignment is enabled by strong, committed project leadership. Such leaders both provide cohesiveness within the team and create support within the organization.

### Staffing

As previously noted, it is crucial that the right types of information (data), expertise, and knowledge are brought to bear at the right times in the process. Availability and utilization of staff are critical in this regard. This is where the corporate and project organizational interfaces can either collide or meld. Tailoring the *who, when,* and *how* in staffing for the development process is not an insignificant task. It cuts into the

> Unfortunately, the experience in most large companies today is that
> organizational and cultural barriers are the single greatest deterrent to
> obtaining acceptance of new paradigms such as P2 or GC, let alone to
> achieving success in integrating even one component of EHS into the
> business operations.

territorial and cultural barriers of different organizations. In addition, it must also work with the existing process development systems, assuming that they are functioning properly.

Chapter 3 introduced the types of traditional staffing (or knowledge centers) typical of different stages in process development. Obviously, this must be expanded to include other means of knowledge transfer than personal involvement to support EHS integration. The goal is to allow knowledge transfer to occur in a way that does not overly encumber the development process with masses of people and frequent, lengthy meetings. The complement of disciplines, experts, leaders, and facilitators/coordinators involved at each step along the way, therefore, should be carefully choreographed in advance and adjusted as needed throughout the process. The same is true for the manner in which the communications and interchanges are to occur—meetings, teleconferences, etc. Tools that bring the critical EHS knowledge to the process development team without requiring EHS bodies as well can be very effective.

Despite the desirability of avoiding highly structured organizations, some degree of structure within the project team will inevitably be required unless the projects are very small and short-lived. This can be accomplished through the use of both Core Teams and support groups in terms of staff utilization. These are common in development processes; however, traditional staffing will need to be expanded both horizontally and vertically. This includes the use of staff to accomplish better transitions between stages, effectively actualizing the five fundamental principles.

## 4.2.4. Implementation Elements

The five steps comprising the implementation elements provide a format for conducting EHS evaluations within the construct of the development process. Figure 4-3 presents a simplified diagrammatic representation of the linkage of these five steps. It parallels a similar process that has been fairly widely endorsed for conducting P2 assessments.

The following provides a consensus list of guidelines extracted from personal interviews and workshops with more than a dozen leading practitioners intimately involved in E, H, and S evaluations for process development.

**Do**

Incorporate as much "threshold" level information as possible into tools and make the tools widely available to minimize nonproductive staff time in routine/repetitive information transfer and "uploading"

- Try to involve as many knowledge and technical advocates as reasonably possible in crafting and selecting tools. This not only results in better tools, but also educates and creates acceptance/buy-in
- Use well-recognized, "neutral" facilitators when working with more than two separate, potentially conflicting groups
- Select Project Leaders carefully
- Involve as many multiskilled people as possible
- Advertise the effort to foster a "need to succeed" within the team
- Go outside the company (business unit) to supplement core competencies (fill out perspectives) when required

**Do Not**

- Have huge or long meetings
- Go into the effort without a clear understanding of the objectives and desired outcomes
- Create highly structured organizations

Two of these elements, the first (establishing or refreshing the basis) and last (reconciling issues and decision making), are commonly found in most development processes, particularly those structured using stages with stage-gate criteria. These serve as the "tie-lines" of

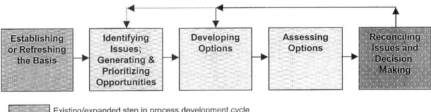

| Establishing or Refreshing the Basis | Identifying Issues; Generating & Prioritizing Opportunities | Developing Options | Assessing Options | Reconciling Issues and Decision Making |

Existing/expanded step in process development cycle
New/adjusted steps to integrate EHS issues
Stage gate in process development cycle

*Figure 4-3. The "Structure" of the implementation elements.*

MERITT to existing development processes. They will probably require some degree of expansion or adjustment to accommodate the integration of EHS with the existing process. The middle three elements may or may not exist in some form within a company's existing development processes.

Following the exact format laid out for these five elements is certainly not necessary; nevertheless, it is important that the content of these elements be covered in some fashion. For example, in small projects with well-defined constraints, it may be possible to combine developing and assessing options or, perhaps, assessing options and reconciling issues. However, short circuiting steps or excluding the content of elements altogether may lead to missed opportunities and additional unplanned work at the end.

1. *Establishing (or Refreshing) the Basis*—The importance of taking the time to carefully and completely establish the basis for what is to be done cannot be overemphasized. Most development processes establish a basis in some form at the outset and carry it throughout the development effort. It should contain all the requirements needed to define both the starting and end points. For most process development efforts, it is rarely known in the earliest stages what the final design, product or plant will actually look like; however, it is possible to define at points along the way what it is to be (or not be) and what it is to do (or not do). Consequently, the basis will normally change as the process development progresses, and those changes may be significant.

   The constituents of the basis can be conveniently grouped according to functionality as shown in Table 4-1. These groupings are but one example and are not intended to be prescriptive in any way. They are intended only to illustrate the process design aspects that typically need to be addressed, as well as how a basis can evolve through development stages.

2. *Identifying Issues (and Generating and Prioritizing Opportunities)*—Issues tend to be specific to different stages of development, at least with respect to their sharpness of focus. At the earliest stages in development of a new process, they are usually fairly broad and relate more to overall process attributes or

TABLE 4-1. Illustration of the Evolution of a Process Development Basis

| Development Stage | Concept Initiation | Basic Engineering |
|---|---|---|
| **Objectives** | | |
| General Process | Product Quality | Production Capacity |
| | Product Efficacy | Plant Availability/Reliability |
| | Synthesis Scale-up | Plant Investment Cost |
| EHS | No VOHAPs | Below NESHAPs Limits |
| | No Chlorinated Solvents | No Major Source Categories |
| | | Minimize Inventories |
| **Criteria** | | |
| General Process | Process Simplicity | Product Quality |
| | Process Yield | Operational Efficacy |
| | Synthesis Controllability | Process Economics |
| EHS | Limited Number of Solvents | Defined Risk FN Limit |
| | Limited Water Use | Specified Inventory Thresholds |
| | Generation of Toxic Materials on Demand | |
| **Requirements/Constraints** | | |
| General Process | Corporate Process Guidelines | Corporate Design Standards |
| | Product Commercialization Date | Plant Commercialization Date |
| | | Utilities Availability/Cost |
| EHS | No Highly Toxic Materials Used | Environmental Regulations |
| | Low Temperature/Pressure Processes | OSHA Regulations |
| | Recycle/Reuse | Community Acceptance |

requirements. In later stages, as the process design takes shape, they can become quite specific and often relate to particular performance aspects. For example, in developing a process design for a western U.S. plant site, water is frequently identified as a major environmental issue both in terms of water consumption and the quality of wastewater produced. In the early stages of process development this may be identified as simply minimizing water consumption with a eye toward the core process units.

By the time the design development reaches the Definition or Basic Engineering stage, this may turn into integrating the entire process water balance (either alone or within a larger plant) to eliminate process water discharges entirely.

> Inquiry tools are meant to give guidance in not only identifying issues but, more importantly, also articulating the issues within a proper context, and then supporting how they are prioritized. The tools are intended to foster an integrated perspective in thinking about EHS issues concurrently.

3. *Developing Options*—Many issues often have more than one area of impact—e.g., environment and health, or health and safety. For example, hexavalent chromium is considered both an environmental and a health problem, in whatever media it exists or is discharged—air, water, or solids. Volatile organic hazardous air pollutants can represent impacts to all three areas. In addition, potential options to resolve an issue predominantly in one area can result in impacts on others. This underscores the importance of an integrated EHS perspective in both developing and screening options. It clearly offers the opportunity for enormous efficiencies in the process through an immediate forum for discussion of trade-offs.

   There has already been considerable thinking along these lines. Many of the tools listed in Figure 4-2 already incorporate at least two perspectives in their formulations; a few address all three. The reaction hazards analysis tool focuses on safety but addresses significant components of acute, event-oriented health and environment impacts. The Solvent Selection Guide developed under the auspices of GC addresses all three perspectives. This tool is one of the most advanced tools in current use in terms of integrated EHS thinking, particularly in the United States.

4. *Assessing Options*—The exact approach used in assessing options depends upon the stage of development, the complexity of the situation, and the number of alternatives identified. In earlier stages, this can often be accomplished in brainstorming sessions using thinklists and performance indicator tools. In the later stages of structured, stage-gate processes, this may require several steps of successively more detailed analyses.

In a multistep assessment, the first step is usually screening the options to divide the options into groups according to relative promise. Two or three such groupings are most common. This can be done using a set of prioritized criteria, often called "gates," that represent "pass/fail" type requisites. Typically, anywhere from two to five such criteria are employed, but fewer criteria make it easier to clearly partition the options. The most promising group of options is then ranked according to a second set of criteria, which may be weighted to reflect their relative importance. This usually requires some degree of analysis. The analysis can vary from fairly simplistic bounding indicators of relative merit using performance indices to detailed life-cycle analysis or total cost assessment. In some cases, breaking this analysis into two steps is advantageous in focusing the effort on only the most promising options.

The criteria used in these assessments should be established up front as a part of the basis. It is not necessary that the criteria used for screening or ranking the options be totally different. Indeed, some screening criteria reflect attributes that are not only necessary but are also of significant "relative" importance.

Obtaining information from all perspectives is critical in assessing options. This is where the "rubber meets the road" in EHS integration—that is, the EHS *fusion* concept reflective of concurrency, collaboration, and communication. Achieving this draws upon all of the resource components (criteria, metrics, tools, staffing, and alignment). It is important to recognize that in the early stages, these assessments are generally an exercise in *local optimization*, while in later stages they usually take a broader perspective of the overall process and even the plant.

5. *Reconciling Issues and Decision Making*—Reconciling issues and decision making should follow the basic approach already used for all other aspects of process development. However, in virtually all cases it will be necessary to ensure that the integrated EHS perspective is included in the criteria or metrics used. It may also be necessary to expand the group of participants involved in key decision-making exercises or to provide some training to help decision makers better understand EHS information.

Changing the decision-making approach could be a fatal flaw, in that it could be perceived as a challenge to the responsibilities and authorities of various individuals and it would encounter all sorts of cultural barriers.

## 4.2.5. Overall Format—Fitting It All Together

Figure 4-4 presents an expanded version of the implementation elements diagram shown in more simplified form in Figure 4-3. Figure 4-4 diagrammatically incorporates four of the resource components—criteria, metrics, tools, and staffing. Alignment is not shown since it is implicit to all of the tools and reinforces each and every fundamental principle. Together, these complete the functional format of MERITT and present one integrated EHS approach, but by no means the only one. Whether it be MERITT or some other approach, three attributes contribute importantly to success:

- Credibility
- Value creation
- Flexibility

### Maintaining Credibility

Paramount to the success of EHS integration is maintaining the "credibility" of the overall development process. This relates directly to process viability. This does not mean that EHS performance or criteria must be sacrificed to assure financial viability. In fact, just the opposite is true. EHS integration must work within the development process to ensure that the process design concurrently meets *all* criteria. EHS criteria must be realistic and suitable.

Thus, overall project feasibility must be sustained throughout the process. Five key facets to this success have been identified as economic viability, product quality, operational efficacy, process reliability/maintainability, and EHS performance. Together these constitute a "credit card" for the "transactions" that occur during the development process.

### Value Creation Opportunities—The IS/P2/GC/DfE/GT Interface

What validates the credit card is the value that integrated EHS brings to the final process design. This value accrues from the quality of opportu-

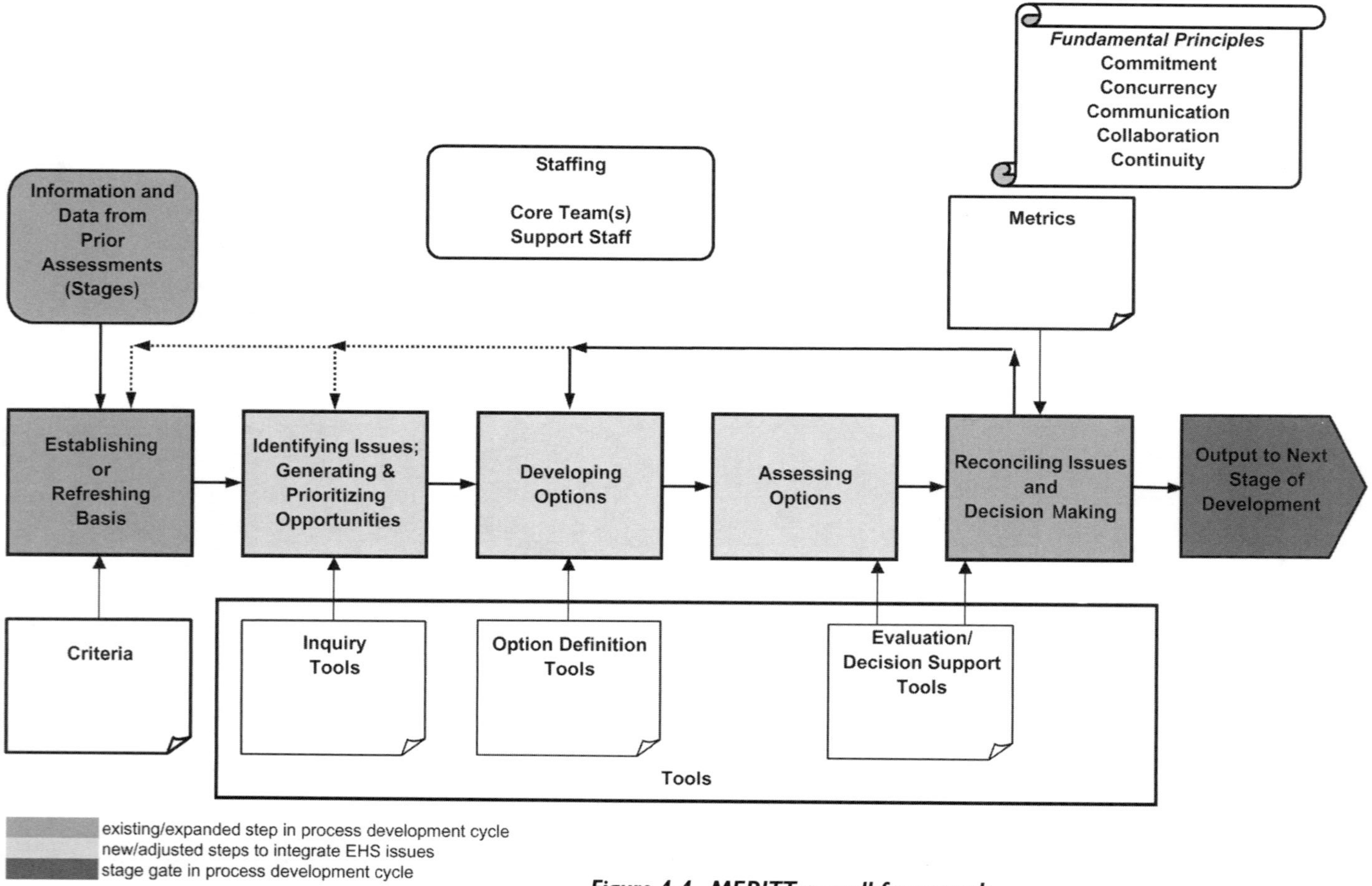

Figure 4-4. MERITT overall framework.

*Fundamental Principles*
Commitment
Concurrency
Communication
Collaboration
Continuity

Metrics

Output to Next Stage of Development

Reconciling Issues and Decision Making

Assessing Options

Evaluation/ Decision Support Tools

Staffing
Core Team(s)
Support Staff

Developing Options

Option Definition Tools

Identifying Issues; Generating & Prioritizing Opportunities

Inquiry Tools

Information and Data from Prior Assessments (Stages)

Establishing or Refreshing Basis

Criteria

Tools

existing/expanded step in process development cycle
new/adjusted steps to integrate EHS issues
stage gate in process development cycle

70

> **Project feasibility depends on**
> - Economic viability
> - Product quality
> - Operational efficacy
> - Process reliability/maintainability
> - EHS performance

nities identified within the appropriate timing of the development process. This is what MERITT is devised to achieve. But what makes it doable is the existence of a variety of concepts already available primarily from GC, P2, IS, and DfE. Figure 4-5 presents an Opportunities Matrix that correlates these concepts with different stages of development, as well as for retrofits/upgrades. These will be further refined within the context of the application of MERITT discussed in Section 4.3.

*Flexibility in Utilization Techniques*

The flexibility inherent in the construction of MERITT offers a multiplicity of techniques for utilization. Three are viewed as the most common scenarios for integrating EHS. First is the direct incorporation of MERITT into existing development processes by adoption and adaptation. Second is the selective extraction of elements for infusion into specific development areas. Finally there are occasions where MERITT is used as a basis for formulating a new development process.

## 4.3. Application of MERITT to Stage-Gate Development Processes

Four general process development stages were identified in Chapter 3 that are of particular importance with regard to EHS integration:

1. Concept initiation
2. Process chemistry
3. Process definition, for three different cases:
   a. New processes
   b. Process replications
   c. Process upgrades
4. Basic process engineering

| Strategy/Tenet | Concepts | Inherent Safety | Pollution Prevention | Green Chemistry | Green Technology | Design for Environment | Concept Initiation | Process Chemistry | Process Development | Basic Process Engineering | Detailed Engineering Design | Application to Retrofits/Upgrades |
|---|---|---|---|---|---|---|---|---|---|---|---|---|
| **Substitution** | Synthesis Route - Reaction Chemistry | ✓ | ✓ | ✓ | ✓ | ✓ | ● | ● | ◐ | | | |
| | Feedstocks & Reagents | ✓ | ✓ | ✓ | | ✓ | ◐ | ● | ◐ | ○ | | ● |
| | Catalysts | ✓ | ✓ | ✓ | ✓ | ✓ | ◐ | ● | ◐ | ○ | | ● |
| | Solvents | ✓ | ✓ | ✓ | ✓ | ✓ | ◐ | ● | ◐ | ◐ | | ● |
| **Minimization** | Process Intensification | ✓ | ✓ | ✓ | ✓ | ✓ | | ○ | ● | ● | ○ | ● |
| | Inventory Reduction | | ✓ | | ✓ | ✓ | | | ○ | ● | ○ | ● |
| | Recycle | ✓ | ✓ | ✓ | ✓ | ✓ | | | | ● | ○ | |
| | Focused Analytical Techniques | ✓ | ✓ | | | ✓ | | | | ● | | |
| | Plant Location ("Co-Location") | ✓ | ✓ | | | ✓ | | | ● | ● | | |
| **Simplification** | System Design (Multi-step vs. Integrated) | ✓ | ✓ | ✓ | ✓ | ✓ | ○ | ◐ | ● | ● | | ● |
| | DCS Configuration | ✓ | ✓ | | ✓ | ✓ | | | ◐ | ◐ | ◐ | |
| | Pre-purified Raw Materials | ✓ | ✓ | ✓ | | ✓ | | ◐ | | ● | ○ | ● |
| | Individual Equipment Design | ✓ | | | | ✓ | ◐ | | ◐ | ◐ | ○ | ● |
| **Moderation (1)** | Conversion Conditions (pH, T, P) | ✓ | ✓ | ✓ | ✓ | ✓ | | ● | ● | ○ | ◐ | ● |
| | Storage Conditions (T, Form, State) | ✓ | ✓ | | ✓ | ✓ | | | ● | ● | | ● |
| | Dilution (Heat Sink, Reaction Kinetics) | ✓ | ✓ | | | | | | ● | ◐ | | |
| | Equipment Overdesign (Pressure) | ✓ | ✓ | | | ✓ | | | ○ | ● | ◐ | |
| **Moderation (2)** | Offsite Reuse | | ✓ | | | ✓ | | | | ● | | ◐ |
| | Advanced Waste Treatment | | ✓ | | | ✓ | | | ○ | ● | ○ | ◐ |
| | Benefical Co-Disposal | | ✓ | | | ✓ | | | | ◐ | ○ | |
| | Equipment/Process Cleaning Design | ✓ | ✓ | | ✓ | ✓ | | | ◐ | ● | | |
| | Plant Cleanup Practices (Dry vs. Wet) | | ✓ | | ✓ | ✓ | | | | ◐ | | ○ |
| | Plant Location (Climate) | ✓ | ✓ | | | ✓ | | | ● | ◐ | ◐ | |
| | Monitoring Systems | ✓ | | | | | | | ○ | ● | ● | |
| | Secondary Containment | ✓ | | | | | | | ○ | ● | ● | |
| | Backup/Redundant Systems | ✓ | | | | | | | | ◐ | ● | |
| **Energy Efficiency** | Waste Heat Recovery (Cascading) | | ✓ | | ✓ | ✓ | | | ○ | ● | ◐ | ◐ |
| | Fuel Mix | | ✓ | | ✓ | ✓ | | | ○ | ● | ◐ | |
| | Heat Transfer Equipment Efficiency | ✓ | | | | | | | ○ | ◐ | ● | ● |

Legend:
- ✓ Basic Concept
- ⟋ Related Concept
- ● High Potential
- ◐ Moderate Potential
- ○ Low Potential

*Figure 4-5. Value creation opportunities matrix.*

The delineation among these stages reflects typical industry practices and nomenclature, but are, to an extent, somewhat arbitrary. The scope of activities in any one stage will vary by industry, by company, and even within companies according to business units and the types of development processes they have specified. Therefore, it is important to note that characteristics attributed to stages and the approach to EHS integration for a particular stage described herein may actually have aspects that relate to either preceding or succeeding stages in company-specific applications.

It is important to note that the very structure of stage-wise development processes can present barriers. Discrete start and end points to stages tend to create discontinuity often due to the hand-off of responsibilities and the (incorrectly) perceived completion of threshold activities.

### 4.3.1. Early Stages–Concept Initiation (Stage 1) and Process Chemistry (Stage 2)

The concept initiation stage as well as the formative deliberations in the process chemistry stage are the defining moments in the process design. They literally "set the stage(s)" for the work to follow. It can then become either a natural complementary extension or an uphill battle to deal with the EHS issues created. Until fairly recently, with the advent of GC and other more company-specific EHS programs, there has been little consideration of integrated EHS, much less focus on any of its components, at these early stages of process development.

There are certainly formidable cultural barriers and institutional disincentives in most large organizations to integrating EHS at these stages. These barriers are difficult to deal with and can take considerable time and effort and are addressed briefly in Chapter 7. There are, however, more tractable issues related to integrating EHS. The fundamental ones are awareness and competency. Few synthesis or development chemists are aware of the full consequences of their decisions on process EHS performance; and they are rarely schooled in the basics of reaction scale-up, process engineering, and manufacturing practicalities. Exposure to EHS issues is further limited by both natural aversion to "outsider" intrusions and the difficulty in gaining access to knowledgeable EHS practitioners and process designers.

Advancing EHS integration to the overall development process, therefore, should focus on four facets:

1. *Training* needs to generate a degree of awareness which leads not just to an environment of acceptance but to a more proactive appeal for support. This requires a demonstration of the overall business value of front-end loaded EHS integration; and, closely associated with this, the value to individual professionals in making it happen.

2. *Manageable Opportunities* are essential to success. In these early stages, the focus should be on what is within the scope and context of the development effort. Examples include:

   • Types of reactions
   • Yields
   • Separations
   • Types of synthesis steps (operations)
   • Byproducts
   • Feedstocks and reagents
   • Solvents
   • Catalysts
   • Conversion conditions (T, P, pH, concentrations)

   Attention should be paid not only to EHS issues intrinsic to the chemicals used and produced, but also to the processing requirements that are inherent to the synthesis route itself (e.g., chlorinated organics, azeotropes). In addition to investigating opportunities for enhancing EHS performance, information must also be developed for use in process optimization in succeeding development stages.

3. *Tools* need to support "segmental optimization"—specifically optimization that addresses ingredients, reactions, and process steps. At these early stages there is usually insufficient information about the overall process and plant application to allow broader-based optimization considerations. However, it is possible to minimize "downstream" development problems by utilizing tools with built-in projections of impact potential, particularly thinklists and selection guides that can be used by individuals or small groups. These tools can be particularly critical in fast track development situations. Examples of such tools include:

- Chemical Hazards Analysis
- Chemical Interaction Matrix
- HAZOP
- Reaction Selection Guide (under development)
- Solvent Selection Guide

4. *Criteria and Metrics* should be introduced that present at least a coordinated EHS perspective if not a joint one. They must be explicit, but should be simple and somewhat flexible. Product and process efficacy are still the primary goals at this stage and EHS criteria should not interfere with the natural flow of development.

E, H, and S practitioners and process design experts should be involved, but as a supporting cast. They cannot lead, nor should they facilitate, the development efforts. Their roles should be to give guidance, information, and advice, and to provide a context to the development effort. They should also participate actively in reviews.

## 4.3.2. Process Definition (Stage 3)

Once the basic concept has been identified, overall optimization of the core process is usually undertaken. Attention now shifts to finalizing the process technologies to be used and defining the basic operating parameters for the unit operations. The principal process equipment is also finalized. This usually entails pilot scale or possibly prototype testing supported by process simulation.

At this point, a more diverse and interactive development team is needed, led by a process development engineer. By incorporating EHS perspectives on the development team, often in person, some potential problematic EHS issues can be avoided while it is still cost effective and fairly easy to do so, including:

- Scale-up problems associated with raw material transport, storage, and handling requirements
- Equipment that is hazardous or difficult to operate and/or maintain
- Occupational safety issues associated with certain process equipment or activities (such as manual catalyst additions or cleaning filter presses)
- Excessive waste generation or water usage

The emphasis at this stage is therefore on broadening the knowledge and experience brought to the design, primarily through the participation of EHS professionals or increased training for various engineers involved in process development. Keeping the chemists involved through this stage to help identify and evaluate alternatives can be critical. Participation need not be full time, if that is the route chosen.

Criteria also become important, as many alternatives may need to be developed and reviewed in a fairly short period of time.

### 4.3.3. Basic Process Engineering (Stage 4)

The basic process engineering stage is one where E, H, and S perspectives are often considered today, albeit individually, at different times and to varying extents. Under MERITT, the only significant changes foreseen are the more integrated and continuous involvement of EHS perspectives and the inclusion of EHS criteria along with the other stage-gate criteria.

## 4.4. Fast Tracking with MERITT

The impact of fast tracking on a typical staged development process is compression of time (schedule), which in turn affects how stage activities are carried out and coordinated. Therefore, a fast-track process development can possess many of the following attributes:

- Shorter schedules
- Overlapping stages
- Stage short-circuiting
- Combining stages
- Parallel activities/evaluations
- Parallel decision making
- Independent decision making
- Local optimization, which may be time dependent and/or development event driven

These same attributes may also be found in small companies and in other situations where a lot of detail is not needed. The effects will vary depending upon the degree of schedule compression. In more moderate cases, the result may be some amount of stage overlap or stage short-circuiting, such that the decision to start the next stage occurs before the present stage is completed. In the extreme, discrete stages may dissolve entirely leaving a succession of parallel and overlapping activities with independent decision making at the activity level. Consequently, as the compression increases, project management moves farther and farther away from "orchestration" and more and more toward "herding." In these situations, joint or established approaches tend to channel the development decisions, at least at the local level, enhancing the credibility of and incentive for a truly integrated EHS approach. A well thought out, integrated EHS approach should inherently be more efficient and faster than the more traditional, independent, and iterative sequencing of environmental, health, and safety activities. The advantage gained is much better closure on a design while achieving the stated objectives and criteria, with no sacrifice in terms of the schedule.

It is highly desirable, therefore, to formulate an integrated EHS approach that is capable of meeting the challenges of fast-track development. In terms of MERITT, this usually entails arranging implementation elements to overcome or reduce the constraints, and selecting and configuring resource components to allow more expeditious access to information and more effective decision making. The term *mining* is used to describe this activity. It is, perhaps, an imperfect descriptor; however, it does convey the basic message of extracting what is needed and assembling it in a form that best fits the specific project requirements. Chapter 6 provides an example of how MERITT can be used, drawing upon these concepts.

## 4.4.1. Fast Tracking Implementation Elements

For fast-track projects, the process must be streamlined by judiciously collapsing elements with related or linked activities. An obvious challenge due to schedule compression is that the time interval between key decision points is far shorter, with less time to identify and evaluate project options. Thus, decisions may be distributed rather than centralized in a fully structured stage-gate development process.

Getting the right start is fundamentally important. This means establishing a solid basis and plan of action. Key concepts include:

- Ensuring a well-defined starting point with clearly articulated priorities and criteria;

- Clearly defining the endpoints or outcomes, being realistic about what can and cannot be accomplished, given the schedule and resources available;
- Recognizing the importance of anticipating needs, not only information needs but also issues, to the extent possible; and
- Resolving identifiable implementation issues at the outset.

Next, focus on managing the scope of the process. This usually entails controlling the issues through focus, focus, and more focus—how they are identified, scoped, prioritized, validated, and resolved. At the outset, brainstorm the process to determine what may be the biggest issues, then prioritize them. Two approaches may prove useful in this endeavor.

1. Reduce the "dimensions" of EHS issues if at all possible. This may take recasting or constraining the issues in some manner. The point is to get to the core of each major issue and to internally prioritize each one, as much as possible. Compromises and trade-offs will undoubtedly need to be made here, but it is important to recognize the value of local optimization rather than try to address the broader aspects of all issues.
2. When structuring issues consider both the priority and ability to resolve (see Figure 4-6). Try to develop areas of concentration so that the resources can be appropriately focused. It is of little value to spend inordinate amounts of time and energy trying to resolve issues when full resolution is beyond the time and resources available. Attack the highest priority issues that can be resolved and try to recast others to define an approach which resolves them. Draw upon experiences from similar process designs to the greatest extent possible.

### 4.4.2. Fast Tracking Resource Components

The nature of fast-track projects necessitates a "stripped down" version of resource components.

1. *Criteria/Metrics*—Make them crisp; keep them simple and few; and prioritize them. Consider the customer's criteria and bounds. Set up a trapping system to detect issues, needs or aspects that are "show-stoppers" that will delay or deny project success.

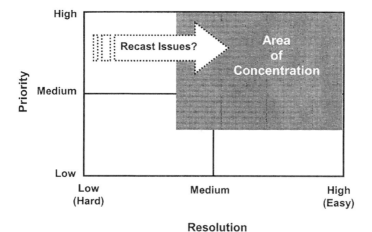

**Figure 4-6. Issues priority/resolution diagram.**

2. *Staffing*—Use experienced, knowledgeable staff with a demonstrated ability to accept compromise and go forward. Experience in similar process designs and fast-track projects is most helpful.
3. *Tools*—Bound problems. Stay away from details as much as possible until absolutely necessary. Use bounding methods in evaluations and decision making, and go after the big tickets only (or at least first). The selection of tools for these types of projects needs to reflect this.

# 5

# MERITT Tools

This chapter discusses various tools that can be used in combination with the MERITT methodology to facilitate the application and integration of EHS tenets. The chapter begins with a discussion of the purpose of tools and broadly defines different types. Next, the relationship among tools, development stages, and applicability is delineated. Information on existing tools and how they support MERITT is provided. Since the MERITT concept is relatively new, many of the existing tools are single discipline in their focus. A few examples are provided on how some of these can be modified to address the collaborative spirit of MERITT.

## 5.1. Introduction

### 5.1.1. Need for Tools

During the initial phase of formulating the MERITT methodology, it became apparent that many companies applied a number of the concepts of MERITT; however, there was little formality or structure to their implementation. Interestingly, the *IN*herent *S*HE *In* *De*sign (INSIDE) Project, an earlier effort undertaken in Europe, arrived at the same conclusion.

Thus, tools are crucial because they are the means by which the concepts of MERITT are put into practice.

> The need for some tools or methods to help consider inherent SHE was a clear message from all the discussions and so the project went on to put forward and develop a number of preliminary ideas for tools that could help process developers and designers identify and consider inherently safer options at the early stages of design.
>
> —*INSET project manager [2]*

## 5.1.2. Purpose of Tools

MERITT tools are designed to help chemists, scientists, and engineers address EHS issues in an integrated way, during the life-cycle stages involving process selection, development, design, and implementation. The tools are intended to supplement or combine with a company's own procedures and processes employed to address IS/P2/GC and similar considerations. Tools are one way of making EHS knowledge and experience available to projects without adding more staff to existing teams.

## 5.1.3. Types of Tools

Surveys of chemical industry companies [1,2] have shown that three types of MERITT tools are required. These include inquiry, option generation, and option evaluation/decision support.

1. *Inquiry (I)*—These tools are used to coordinate and facilitate investigations and group discussions/evaluations as well as individual thinking. Included in this category are also those tools that can be utilized to examine, massage, and scope issues. Examples include thinklists, brainstorming, and the Chemical Interaction Matrix. Each of these, as well as other inquiry tools discussed later, offers a somewhat different approach. Suitability is usually dependent upon organizational structure, project size, the development process and stage of development, and business culture. In the examples given above, thinklists are usually most appropriate for early stages of development processes and are often prepared for use by individuals and small groups in identifying and scoping issues; providing general direction in avoiding the obvious problems; and alerting reviewers to the existence and magnitude of potential areas of concern. Brainstorming, on the other hand, is typically a facilitated, relatively informal (although not totally unstructured) interaction involving small- to medium-size groups of individuals representing related or diverse expertise. Additionally, the Chemical Interaction Matrix requires a fairly intensive, structured thinking process of experts in a field working both individually and together in sequential interactions.

2. *Option Generation (OG)*—These tools are geared toward helping to identify options for minimizing environmental, safety, or health impacts. There are a few tools that may be of particular applicability to integrated EHS, such as: the Process Options Generation Tool (Tool Set B) from INSET; the Solvent Selection Guide; and P2 strategies matrices. Some tools are capable of defining very explicit options (e.g., the Solvent Selection Guide) while others provide broader guidance in channeling thinking (e.g., P2 strategies matrices). Notable by absence are tools specifically focusing on health impacts. Health impacts tend to be subsumed within tools designed to address environmental or safety concerns (e.g., the Solvent Selection Guide).

3. *Option Evaluation/Decision Support (E/DS)*—These tools focus on: helping to choose among alternatives; evaluations against established criteria; and as one part of an overall decision-making process. These tools can be very similar to Option Definition Tools, especially where Option Generation tools are very specific in nature. The tools generally found in this category are: indices (e.g., the Dow/Mond Index and the Reaction Hazards Index); cost analyses, which tend to be somewhat generic and most commonly used for environmental assessments [e.g., total cost assessment (TCA) and life-cycle analysis (LCA)]; and equipment/operations comparison guides [e.g., Unit Ops Comparison Guide (INSET Tool Set O)]. Of particular note is the Inherent SHE Performance Index (INSET Tool Set I) developed specifically for integrated E, H, and S evaluations.

   One method that has been found to be quite useful in comparisons of process designs in general is a modified Kepner-Tregoe (K-T) analysis, where criteria are arranged into groupings such as economics, environmental, health, safety, operability, maintainability, etc. Weightings are then assigned to the groupings and scores are developed for the individual criteria constituents within each group relative to falling short of, meeting, or exceeding the criteria. This allows the contributions of the individual criteria to be fully recognized, which is particularly important where there is little value to be gained by exceeding or far exceeding a threshold criterion.

One of the issues in using nonintegrated E/DS tools is that they utilize different rating or scoring systems. K-T analysis has been found useful in creating a common framework for ranking options, where different evaluation schemes were used for IS and P2.

Examples of each type of tool are listed in Table 5-1. Brief descriptions of various tools are presented in Section 5.2. It is readily apparent that most tools have characteristics that fit more than one category.

## 5.1.4. Relationship of Tools and Stages

An understanding of the relationship of development stages and tools begins with a review of the alignment of EHS strategies with stages. This alignment is shown in Table 5-2. These strategies are available to both new and retrofit projects depending on which stages apply. For retrofit projects, the practitioners may need to be more assertive during the project scoping activities, so that process development and pre-development stages are considered and included when feasible, along with basic engineering.

Tools generally attempt to capture some or all of these strategies, however, there currently are few that truly achieve the integration of all. Section 5.3.2 presents some examples of how certain existing tools can be modified to achieve a more integrated version.

**TABLE 5-1.** *Available Tools*

| Tool Type | Examples |
|---|---|
| Inquiry (I) | Chemical Interaction Matrix, Guide Word Techniques (e.g., HAZOP and FMEA) for Process or Chemical Hazard Analysis |
| Option Generation (OG) | Responses to Guide Word Techniques, Solvent Selection Guide, INSET Tool B |
| Evaluation/Decision Support (E/DS) | Environmental Indices, Layer of Protection Analysis (LOPA), Total Cost Assessment, Engineering Evaluations, INSET Tool J |

**TABLE 5-2.** *Alignment of EHS Disciplines to Project Stages*

| Stage | Green Chemistry | Pollution Prevention | Inherent Safety | Green Technology |
|---|---|---|---|---|
| Concept Initiation | Eliminate, Substitute, Low Persistence, or Bioaccumulation | Eliminate | Substitute | Eliminate, Substitute, Energy Use |
| Process Chemistry | *Process:* Eliminate, Renewable, Order of Steps, Minimize/Simplify *Chemistry:* Atom Economical, Selectivity, Reduced Toxicity, Mass Efficient | Reduce/ Recycle | Minimize/ Simplify | *Reactors, Mixers:* Minimize, Scale, Simplify, Eliminate (steps, unit operations, etc.), Energy Use *Separations:* Substitute, Eliminate |
| Process Development | Attenuate/Moderate | Re-use | Attenuate/ Moderate | Attenuate, Order of Unit Operations, Combination of Steps |
| Basic Process Engineering | Low Waste or Nonproduct | Treat, Contain | Intensify | Intensify |
| Detailed Engineering | | Reliability/ Redundancy | Mechanical Integrity | Reliability, Redundancy |
| Production | | Dispose, Process Redesign | Mitigate, Process Redesign | Process Redesign |

Some tools are generic and can be applied to all stages (e.g., generic option generation tool). Others are more stage specific (e.g., Solvent Selection Guide). Figure 5-1 summarizes the applicability of many existing tools to the various development stages and indicates the EHS discipline(s) for which they have traditionally been applied. A description of these tools is provided in the next section. As the matrix shows, some tools (e.g., what-if, thinklists) are generally applicable at most any stage, but work best during certain stages depending on the level of information available. Other tools (e.g., Reaction Hazard Analysis, Solvent Selection Guide) were developed specifically to be used during the earlier stages of process creation. Figure 5-1 is provided as a general guide to when various tools would be appropriate for application.

| Tool Type | Tool | Inherent Safety | Pollution Prevention | Green Chemistry | Concept Initiation | Process Chemistry | Process Definition | Basic Process Engineering | Detailed Engineering Design | Applicability to Retrofits/ Upgrades |
|---|---|---|---|---|---|---|---|---|---|---|
| | | | | | Applicability to Stage | | | | | |
| Inquiry | Thinklists | ✓ | ✓ | ✓ | ● | ● | ◐ | | | ◐ |
| | Chemical Interaction Matrix | ✓ | ✓ | ✓ | ● | ● | ● | | ○ | ◐ |
| | Chemical Hazards Analysis | ✓ | ✓ | ✓ | ● | ● | ● | ○ | ○ | ● |
| | Brainstorming | ✓ | ✓ | ✓ | ● | ● | ● | ● | ○ | ● |
| | What-if Analysis | ✓ | ✓ | ✓ | ○ | ○ | ● | ● | ● | ● |
| | Process SHE Analysis* | ✓ | ✓ | | ○ | ○ | ● | ● | ● | ● |
| | Guide Word Process Hazards Analysis | ✓ | | | ○ | ○ | ◐ | ● | ● | ● |
| Option Generation | IS strategies[a] | ✓ | ✓ | | ● | ● | ● | ◐ | ◐ | ● |
| | P2 strategies[b] | ✓ | ✓ | ✓ | ● | ● | ● | ◐ | ○ | ● |
| | Solvent Selection Guide | ✓ | ✓ | ✓ | ● | ● | ○ | ○ | ○ | ● |
| | Reaction Selection Guide | ✓ | ✓ | ✓ | ● | ● | ○ | ○ | ○ | ● |
| | Reaction Hazards Analysis | ✓ | ✓ | | ● | ● | | | | ● |
| | Process Option Generation Tool* | ✓ | ✓ | ✓ | ● | ● | ◐ | ◐ | ◐ | ● |
| | Brainstorming | ✓ | ✓ | ✓ | ● | ● | ● | ● | ● | ● |
| | CCPS Safer Design Solutions | ✓ | | | ○ | ○ | | | | ● |
| | API 752-Siting | ✓ | ✓ | | ● | | ○ | ◐ | ○ | ● |
| | Dow/Mond Index | ✓ | | | ● | ● | ● | ◐ | ● | ● |
| | Inherent Safety Index | ✓ | ✓ | ✓ | ● | ● | ● | ● | ● | ● |
| | Environmental Indices | ✓ | ✓ | | ● | ● | ● | ◐ | ◐ | ● |
| | Health Indices | ✓ | ✓ | ✓ | ● | ○ | ● | ◐ | ◐ | ● |
| | Reaction Hazard Index | ✓ | ✓ | ✓ | ● | ◐ | ● | ○ | ○ | ● |
| Evaluation/ Decision Support | LOPA | ✓ | ✓ | | ○ | ○ | ● | ● | ● | ● |
| | Fault Tree Analysis | ✓ | ✓ | | ○ | ○ | ● | ● | ● | ● |
| | Total Cost Assessment | ✓ | ✓ | ✓ | ◐ | ○ | ● | ● | ● | ● |
| | Life-Cycle Cost | ✓ | ✓ | ✓ | ● | ● | ● | ● | ● | ● |
| | Process Simulation | ✓ | ✓ | | ○ | ○ | ● | ● | ● | ● |
| | Inherent SHE Performance Index* | ✓ | ✓ | ✓ | ○ | ◐ | ● | ● | ● | ◐ |
| | MERGE | | ✓ | | ● | ● | ◐ | ○ | ○ | ● |
| | Mass Integration | ✓ | ✓ | | ○ | ○ | | | | ◐ |
| | Unit Ops Comparison Guide* | ✓ | ✓ | | ○ | | ● | ● | ● | ◐ |
| | Equipment Simplification Guide* | ✓ | ✓ | | ○ | ○ | ◐ | ● | ● | ● |

\* Tools from INSIDE Project

[a] Minimization, Moderation, Substitution, Simplification

[b] Minimization, Recycle/Reuse, Alternative Waste Treatment

✓ Typical Use
✓ Also Useful

● High Applicability
◐ Moderate Applicability
○ Seldom Applied

**Figure 5-1. MERITT tools applicability matrix.**

## 5.2. Available Tools

The majority of the current tools that are applicable to MERITT have been developed in the context of the specific subpractices of MERITT–namely, inherent safety, pollution prevention, and, to a lesser extent, green chemistry–rather than for an integrated methodology. There are a few tools that were specifically developed to address a MERITT-style methodology. In particular, the Solvent Selection Guide incorporates integrated EHS considerations into the early development stage decisions. Another example is the Inherent SHE Tools (INSET) Toolkit developed by the INSIDE Project [1] that contains tools that support Integrated SHE in Design (INSIDE).

### 5.2.1. Nonintegrated Tools

Most of the existing tools that might be considered for use with the MERITT approach were developed for a single EHS discipline or practice area. For example, the Hazard and Operability (HAZOP) study was initially developed to identify process safety issues that need to be addressed by the project design team. Likewise, there are various environmental indices that were developed to evaluate design alternatives based on environmental worthiness.

The fact that some of these were originally single purpose tools does not necessarily limit their use to a single practice area. As Figure 5-1 demonstrates, many of the tools can and have been utilized for some or all of the EHS disciplines. Furthermore, by appropriate modification (see Section 5.3.2), some of these tools can be transformed into a truly integrated MERITT tool.

A brief description of the more widely used tools is provided below. Refer to Table 5-1 for the definition of tool type categories.

| Tool | Type | | | Applicability | | Reference |
|------|------|---|---|---------------|---|-----------|
| Chemical Interaction Matrix | I | OG | E/DS | New | Retrofit | Most companies have their own |
| | | | | | | |

*Description:* A matrix composed of a list of all chemicals available at the proposed manufacturing unit displayed on each axis. The intersection of the various chemicals is reviewed to determine whether there are any reactions possible that can present a hazardous condition.

| Tool | Type | | | Applicability | | Reference |
|------|------|------|------|------|------|------|
| Hazard and Operability Study (HAZOP) | I | OG | E/DS | New | Retrofit | [9] plus CIA |

*Description:* A process deviation analysis that utilizes a set of standard guide words to assist a skilled team in the identification of hazards and operability issues. The HAZOP technique is generally the tool of choice for process hazards analysis (PHA).

| Tool | Type | | | Applicability | | Reference |
|------|------|------|------|------|------|------|
| Chemical Hazards Analysis | I | OG | E/DS | New | Retrofit | [26] |

*Description:* A variation of HAZOP that utilizes the standard guide words in combination with modified parameters to analyze chemical and reaction hazards.

| Tool | Type | | | Applicability | | Reference |
|------|------|------|------|------|------|------|
| Team Brainstorming | I | OG | E/DS | New | Retrofit | [8] |

*Description:* Brainstorming uses the synergy of group dynamics, coupled with the ingenuity of each participant, to generate a set of ideas superior to what could have been achieved by each person working alone.

| Tool | Type | | | Applicability | | Reference |
|------|------|------|------|------|------|------|
| Team Brainwriting | I | OG | E/DS | New | Retrofit | [8] |

*Description:* Similar to brainstorming except team members fill out brainwriting sheets with ideas. Each member puts down two ideas and then places the sheet back in the center of the table and takes another sheet. This technique permits anonymous input of ideas.

| Tool | Type | | | Applicability | | Reference |
|---|---|---|---|---|---|---|
| What-If Analysis | I | OG | E/DS | New | Retrofit | [9, 10] |

*Description:* What-If Analysis is a form of team brainstorming during which participants are encouraged to vocalize What-If questions or issues of concern. The technique can be applied to almost any type of system or procedure. The technique is sometimes implemented with the aid of an initial set (checklist) of What-If questions.

| Tool | Type | | | Applicability | | Reference |
|---|---|---|---|---|---|---|
| Thinklists Analysis | I | OG | E/DS | New | Retrofit | [9] |

*Description:* Thinklists (checklist) Analysis involves the use of a list of specific items to identify known types of hazards, design deficiencies, and potential accident situations associated with common process equipment and operations. This analysis is knowledge based and typically used to evaluate a specific system design, but is also used in early stages of new process creation to eliminate safety and environmental hazards identified through years of operation of similar systems.

| Tool | Type | | | Applicability | | Reference |
|---|---|---|---|---|---|---|
| Inherently Safer Strategies (Minimization, Moderation, Substitution, Simplification) | I | OG | E/DS | New | Retrofit | [11] |

*Description:* While there is no generally applied tool for addressing inherently safer design, the four main concepts are often applied when using or responding to issues found with other hazard analysis techniques such as HAZOP, Failure Mode and Effects Analysis (FMEA), or What-If. The reference describes the concepts of inherently safer design and contains information on IS review methods and checklists.

| Tool | Type | | | Applicability | | Reference |
|---|---|---|---|---|---|---|
| Pollution Prevention Strategies (Minimization, Recycle/Reuse, Alternative Waste Treatment) | I | OG | E/DS | New | Retrofit | [12] |

*Description:* Like IS, P2 strategies do not rely on any generally accepted tool. The reference contains information on how to develop and implement a pollution prevention program and describes preferred pollution-prevention engineering technologies and practices for reducing waste generation and emissions.

| Tool | Type | | | Applicability | | Reference |
|------|------|------|------|------|------|------|
| Solvent Selection Guide | I | OG | E/DS | New | Retrofit | [3] |

*Description:* A tool for ranking solvents according to their inherent environmental, health, and safety hazards which utilizes a scoring scheme based on four categories of factors including waste, impact, health, and safety. See Section 5.2.2 for a more complete description.

| Tool | Type | | | Applicability | | Reference |
|------|------|------|------|------|------|------|
| Process Options Generation (INSET Tool B) | I | OG | E/DS | New | Retrofit | [1, 2] |

*Description:* Provides both guide word prompt lists and a thinklist (for nonflowsheet representations) to motivate creative thinking on inherent EHS strategies. The purpose is to rigorously challenge route and process alternatives in order to obtain a more inherent EHS process.

| Tool | Type | | | Applicability | | Reference |
|------|------|------|------|------|------|------|
| Design Solutions for Process Equipment Failures | I | OG | E/DS | New | Retrofit | [13] |

*Description:* This guideline book is concerned with engineering design that reduces equipment failure risk due to process hazards. The book identifies the available design solutions that might avoid or mitigate the failure in a series of options that begin with inherently safer/passive solutions.

| Tool | Type | | | Applicability | | Reference |
|------|------|------|------|------|------|------|
| Building Siting in Process Plants API RP 752 | I | OG | E/DS | New | Retrofit | [23] |

*Description:* A risk-based approach for identifying hazards that may affect process plant buildings. The steps involved include building hazard identification, building evaluation (including consequence analysis), and risk management.

| Tool | Type | | Applicability | | Reference |
|---|---|---|---|---|---|
| Dow or Mond Index | I | OG | E/DS | New | Retrofit | [6, 20] |
| | | | | | | |

*Description:* Established and widely applied indices for evaluation of realistic fire, explosion, toxic exposure, and reactivity potential of process equipment and its contents, which utilize the inherent chemical properties of process materials.

| Tool | Type | | Applicability | | Reference |
|---|---|---|---|---|---|
| Inherent Safety Index | I | OG | E/DS | New | Retrofit | [18, 19] |
| | | | | | | |

*Description:* The construction of an IS index using seven parameters (inventory, flammability, explosiveness, toxicity, T, P, and yield) is proposed. Range scores are established for parameter and then used to determine a Chemical Score (first four parameters) and a Process Score (other three parameters), and finally an overall score (CS + PS).

| Tool | Type | | Applicability | | Reference |
|---|---|---|---|---|---|
| Environmental Index | I | OG | E/DS | New | Retrofit | [17] |
| | | | | | | |

*Description:* Index based on acute toxicity data (LC50 and LD50).
- Sums two indices, each measuring the hazard associated with the release into two environmental media:
    1. The aquatic ecosystem, and
    2. The terrestrial ecosystem through food and water intake
- Uses a factor PEC (predicted environmental concentration of the chemical) calculated by Mackay's fugacity model, which gives equilibrium concentration of a chemical in the different environmental compartments

- Factors in the inventory of the material(s)
- Appears sufficiently good for MERITT purposes, practical, easy to use, and fast

| Tool | Type | | | Applicability | | Reference |
|------|------|---|---|---------------|---|-----------|
| Toxicity Hazard Index | I | OG | E/DS | New | Retrofit | [15] |
| | | | | | | |

*Description:* An index used early in process development based on acute toxic effect on humans from inhalation.

- Takes into account the dominant toxic material within the process
- Requires estimation of maximum rate of vapor generation
- Requires using a value for the limiting vapor concentration (emergency response planning guidelines, emergency exposure guidance levels, etc.)
- Allocates penalty factors for material, process, and layout features

| Tool | Type | | | Applicability | | Reference |
|------|------|---|---|---------------|---|-----------|
| Reaction Hazard Index | I | OG | E/DS | New | Retrofit | [16] |
| | | | | | | |

*Description:* An index that provides an estimation of the risk of occurrence of runaway reaction, but not the magnitude of the consequences.

- Based on the exothermic level of the reaction
- Provides penalties for risk factors (for example, biggest penalty is for batch process)
- Provides credits for safety factors (for example, reliable stirring system)
- Appears to be very simple to use

| Tool | Type | | | Applicability | | Reference |
|------|------|---|---|---------------|---|-----------|
| Engineering Evaluations | I | OG | E/DS | New | Retrofit | [12] |
| | | | | | | |

*Description:* Engineering evaluation is the application of a full range of engineering skills to business decision making. It aids decision making by translating technical options into their economic impact—guidance that is fundamental to business decisions. It simultaneously answers the questions: "What is technically feasible?" and "What is it worth?"

| Tool | Type | | | Applicability | | Reference |
|------|------|---|---|---------------|---|-----------|
| MERGE | I | OG | E/DS | New | Retrofit | Alliance for Environmental Innovation (www.edfpewalliance.org) |

*Description:* MERGE is a real-time decision support tool that allows environmental issues to be characterized and compared quickly and easily in the design of formulated products and packaging. It is a Windows-based software application that uses basic product information to quickly generate environmental profiles for product formulations and packing. The tool is designed for staff involved in product design who are not expert in the areas of environmental impact and regulations. Some desirable aspects of the tool:

- Extensive chemical properties database
- Integrates all environmental aspects (including energy intensity)
- Allows what-if analysis
- Applicable at different life-cycle stages
- No environmental knowledge required (embedded in tool)

| Tool | Type | | | Applicability | | Reference |
|------|------|---|---|---------------|---|-----------|
| Mass Integration | I | OG | E/DS | New | Retrofit | [25] |

*Description:* One of the main branches of process integration; mass integration is a systematic methodology that provides a fundamental understanding of the global flow of mass within the process and employs a holistic view in identifying performance targets and optimizing the generation and routing of species throughout the process. Mass integration deals with mass allocation, which is central to pollution prevention. It employs the concept of mass-exchange networks coupled with mass separation agents. Mass integration utilizes powerful tools for determining optimized solutions to process integration problems involving multiple alternatives, and has particular applicability to P2.

| Tool | Type | | | Applicability | | Reference |
|------|------|---|---|---------------|---|-----------|
| Layer of Protection Analysis (LOPA) | I | OG | E/DS | New | Retrofit | [22] |

*Description:* LOPA is a formal risk assessment technique for determining whether adequate levels of protection are provided to manage the risks of process hazards. While less detailed than fault tree analysis, it utilizes many of the same principles.

| Tool | Type | | | Applicability | | Reference |
|------|------|---|---|---------------|---|-----------|
| Fault Tree Analysis | I | OG | E/DS | New | Retrofit | [9] |

*Description:* A structured approach, using logic diagrams, to displaying the different combinations or pathways of various events and failures that can result in a particular undesired event. Can be quantified and used to show the benefits of design changes and mitigation strategies.

| Tool | Type | | | Applicability | | Reference |
|------|------|---|---|---------------|---|-----------|
| Total Cost Assessment | I | OG | E/DS | New | Retrofit | [21] |

*Description:* Total cost assessment is a tool designed to quantify all environmental and health costs, both internal and external, associated with a business decision. The methodology is based on a life-cycle context. Because cost metrics are used, it is readily understood by decision makers. See Section 5.2.2 for a more complete description.

| Tool | Type | | | Applicability | | Reference |
|------|------|---|---|---------------|---|-----------|
| Life-Cycle Cost Analysis | I | OG | E/DS | New | Retrofit | [27] |

*Description:* A methodology to assess the economic (and sometimes environmental) costs of a product at each life-cycle stage, considering: capital investments, operating costs (labor, material, energy requirements), and disposal costs. Uses general accounting principles to account for both current and future costs. Environmental costs can be difficult to estimate.

| Tool | Type | | | Applicability | | Reference |
|------|------|---|---|---------------|---|-----------|
| Inherent SHE Performance Index (INSET Tool I) | I | OG | E/DS | New | Retrofit | [1, 2] |

*Description:* A bundle of index tools that covers all the EHS practice and strategy areas. See the listing of subtools in Appendix B.

A table listing other available tools with reference to descriptive material is provided in Appendix A.

Having tools that are more or less tailored to a single EHS discipline is not necessarily inappropriate. There are practical advantages including:

- Many such tools already exist for different development stages
- They can be applied by the EHS specialists when the timing is appropriate

The obvious disadvantage is when nonintegrated tools are used in a nonintegrated EHS function environment. This can be overcome through integrated teams with diversity in the represented disciplines, assuming this will become part of the normal company operations and culture, which is the ultimate goal.

## 5.2.2. Integrated Tools

At present there is a paucity of integrated MERITT tools. This is to be expected, since the MERITT concept is relatively new. However, there are some specific examples that are worth mentioning. One is a Solvent Selection Guide developed by a member of the pharmaceutical industry [3]. The other is the INSET Toolkit developed in Europe by the INSIDE Project [1,2].

*Solvent Selection Guide*

The Solvent Selection Guide represents an attempt by one company to integrate environment and safety into the fabric of the business and move toward more sustainable business practices while working within its present culture. The prime objective is to provide guidance to chemists for selection of solvents based on their inherent environmental, health, and safety issues, taking into account a variety of general and specific process and facility aspects. The guide is based on an assessment of *key categories* that are considered to be most significant in determining the potential environmental, health, and safety impacts associated with each solvent. For simplification, many different key issues were aggregated into four categories as follows:

- **Waste** (covering incineration, recycling, biotreatment, and VOC)
- **Impact** (air and water)
- **Health** (toxicity and volatility)
- **Safety** (flammability, static, process/chemistry risk)

Each solvent is rated and given a numerical score under the four key categories. The results are summarized in Table 5-3.

A low score (dark shading) does not mean the solvent is absolutely not acceptable, but it is a warning that significant control procedures will need to be provided in the design and that these will most likely adversely impact the cost, and perhaps life-cycle success, of the project. In addition, an overriding premise for use of the tool is that solvents must first be selected to meet the requirements of the chemistry and ultimately product cost through process yield. However, through the use of this tool chemists and development engineers will be more aware of the EHS tradeoffs and therefore can make better solvent selection decisions based on an understanding of the total scope of design issues.

### INSET Tools

The *INSIDE* Project was a major European Community co-funded joint industry project. The objectives of this project were to promote the use of inherent safety, health, and environmental (SHE) protection across Europe and to develop tools to enable chemists and engineers to optimize processes and designs using the "inherent" principles [2]. The INSIDE Project chose to concentrate tool development in the process selection, development, and front-end design stages of projects, as these were considered to be the activities with the greatest opportunities for significant inherent EHS improvement.

A framework for the INSET toolkit was developed based on surveying company needs via interviews. This included the need to identify hazards and other potential problems, prioritize these to select areas where inherent safety efforts could be targeted, help generate EHS alternatives, eval-

## TABLE 5-3. A Guide for the Integration of EHS Factors When Selecting Solvents

Copyright SmithKline Beecham Nov-97

| | Solvent Classification | 1. Incineration | 2. Recycle | 3. Biotreatment | 4. VOC Emissions | 5. Environmental Impact–Aqueous | 6. Environmental Impact–Air | 7. Health Hazard | 8. Exposure Potential | 9. Safety Hazard | Environmental–Waste | Environmental–Impact | Health | Safety |
|---|---|---|---|---|---|---|---|---|---|---|---|---|---|---|
| Acetic Acid (glacial) | 10 | 2 | 2 | 4 | 6 | 8 | 5 | 4 | 4 | 8 | 3 | 6 | 4 | 8 |
| Acetone | 7 | 3 | 4 | 2 | 1 | 7 | 7 | 7 | 5 | 5 | 2 | 7 | 6 | 5 |
| Acetonitrile | 5 | 2 | 3 | 1 | 3 | 8 | 2 | 1 | 3 | 8 | 2 | 4 | 2 | 8 |
| 1-Butanol | 1 | 6 | 2 | 7 | 7 | 8 | 6 | 10 | 7 | 8 | 5 | 7 | 8 | 8 |
| Butyl Acetate | 8 | 7 | 6 | 10 | 6 | 7 | 6 | 7 | 7 | 6 | 7 | 7 | 7 | 6 |
| Diethylene Glycol Monobutyl Ether | 1 | 3 | 4 | 4 | 10 | 9 | 8 | 7 | 10 | 10 | 5 | 8 | 8 | 10 |
| Cyclohexane | 2 | 10 | 4 | 7 | 3 | 4 | 6 | 7 | 6 | 2 | 5 | 5 | 6 | 2 |
| 1,2-Dimethoxyethane | 6 | 3 | 2 | 3 | 4 | 6 | 5 | 4 | 4 | 2 | 3 | 5 | 4 | 2 |
| Dimethyl Acetamide | 5 | 2 | 4 | 3 | 9 | 9 | 8 | 4 | 7 | 9 | 4 | 8 | 5 | 9 |
| Dimethyl Formamide | 5 | 2 | 5 | 2 | 8 | 7 | 8 | 3 | 6 | 7 | 4 | 8 | 4 | 7 |
| Dimethylpropylene Urea | 5 | 2 | 4 | 3 | 10 | 9 | 6 | 4 | 6 | 9 | 4 | 7 | 5 | 9 |
| Ethanol/IMS | 1 | 3 | 2 | 3 | 4 | 10 | 6 | 10 | 8 | 6 | 3 | 7 | 9 | 6 |
| Ethyl Acetate | 8 | 4 | 4 | 5 | 3 | 9 | 8 | 7 | 6 | 4 | 4 | 9 | 7 | 4 |
| Ethylene Glycol | 1 | 2 | 3 | 5 | 10 | 10 | 8 | 7 | 10 | 10 | 4 | 9 | 8 | 10 |
| Heptane | 2 | 10 | 4 | 7 | 4 | 1 | 3 | 4 | 7 | 1 | 6 | 2 | 5 | 1 |
| Hexane | 2 | 10 | 4 | 6 | 2 | 3 | 4 | 4 | 2 | 1 | 5 | 3 | 3 | 1 |
| 2-Propanol | 1 | 3 | 2 | 3 | 4 | 10 | 10 | 7 | 7 | 7 | 3 | 10 | 7 | 7 |
| Isopropyl Acetate | 8 | 5 | 6 | 7 | 4 | 10 | 5 | 7 | 6 | 6 | 5 | 7 | 7 | 6 |
| Diisopropyl Ether | 6 | 9 | 5 | 7 | 2 | 3 | 1 | 7 | 5 | 1 | 5 | 2 | 6 | 1 |
| Methanol | 1 | 2 | 3 | 3 | 2 | 10 | 6 | 4 | 5 | 8 | 3 | 8 | 4 | 8 |
| 2-Methoxyethanol | 1 | 3 | 4 | 4 | 7 | 10 | 8 | 1 | 3 | 7 | 4 | 9 | 2 | 7 |
| Bis(2-methoxyethyl) Ether | 6 | 6 | 4 | 9 | 9 | 6 | 5 | 1 | 4 | 3 | 6 | 5 | 2 | 3 |
| Methyl Acetate | 8 | 3 | 3 | 2 | 1 | 9 | 4 | 7 | 4 | 5 | 2 | 6 | 5 | 5 |
| Dichloromethane | 4 | 1 | 10 | 6 | 1 | 3 | 3 | 1 | 1 | 10 | 3 | 3 | 1 | 10 |
| Methylethyl Ketone | 7 | 4 | 1 | 3 | 3 | 7 | 5 | 4 | 5 | 5 | 3 | 6 | 5 | 5 |
| Methylisobutyl Ketone | 7 | 6 | 7 | 8 | 6 | 10 | 1 | 7 | 6 | 7 | 7 | 4 | 6 | 7 |
| n-Methyl Pyrrolidone | 5 | 2 | 4 | 3 | 10 | 8 | 6 | 6 | 10 | 10 | 4 | 7 | 7 | 10 |
| Petroleum Ether | 2 | 10 | 4 | 5 | 1 | 1 | 3 | 4 | 6 | 1 | 4 | 2 | 5 | 1 |
| Propionic Acid | 10 | 3 | 5 | 4 | 8 | 8 | 8 | 4 | 6 | 9 | 5 | 8 | 5 | 9 |
| Propyl Acetate | 8 | 6 | 8 | 8 | 5 | 8 | 4 | 7 | 7 | 6 | 7 | 6 | 7 | 6 |
| Pyridine | 11 | 2 | 1 | 2 | 6 | 5 | 2 | 1 | 1 | 6 | 2 | 3 | 1 | 6 |
| Methyl t-Butyl Ether | 6 | 6 | 7 | 4 | 1 | 3 | 5 | 7 | 1 | 3 | 4 | 4 | 3 | 3 |
| Tetrahydrofuran | 6 | 4 | 2 | 1 | 2 | 7 | 8 | 4 | 4 | 2 | 2 | 7 | 4 | 2 |
| Toluene | 3 | 10 | 5 | 0 | 5 | 7 | 2 | 4 | 5 | 4 | 7 | 3 | 4 | 4 |
| p-Xylene | 3 | 10 | 6 | 10 | 7 | 5 | 4 | 4 | 7 | 5 | 8 | 4 | 5 | 5 |

Composite scores are given for the four key areas. They are based on a scoring range of 1-10. The higher the score, the better.

- **Waste** addresses recycling, incineration, VOC, and biotreatment issues.
- **Impact** addresses fate and effects on the environment.
- **Health** is based on acute and chronic effects on human health and exposure potential.
- **Safety** considers explosivity and operational hazards.

| 1-3 | Major issues have been identified. Appropriate control procedures need to be in place. |
|---|---|
| 4-7 | Issues have been identified. The need for control procedures should be considered. |
| 8-10 | No major issues have been identified in this area. |

**HAZARD/PROBLEM IDENTIFIER/PRIORITIZER**

• uses existing company datasheets/hazard studies

• hazard/problem record to track hazards

• hazard/problem ranking/prioritizer

**OPTION GENERATOR**

• sets structure for analysis

• sets objectives

• guide word/brainstorm methods

—prompt deviations

—question functionality

—prompt different means to achieve same function

**INITIAL SCREENING**

• compares options against key success factors

• rapid screening to find best options

• warns of possible conflicts between S, H, and E

**DECISION AIDS**

• used where no clear best option identified

• ranking index for inherent S, H, and E

• multiattribute analysis to aid decision making

• defines "musts" and "wants" criteria/constraints

• includes provision for cost, feasibility, and other decision criteria

• provides stand-alone decision support tool or can feed in to existing company decision support tools

*Figure 5-2. Inherent SHE tool framework.*

uate these, and then feed this information into an overall decision framework where required. This framework is shown in Figure 5-2.

The toolkit was developed so that it can be used in a wide variety of situations in the chemical and process industry, depending of the nature of the project concerned and the time at which the toolkit is introduced. As a result, quite a number of tools were developed. A listing of all the tools and their primary aim is shown in Appendix B. The toolkit contains tools that address the three main categories: inquiry, option generation, and option evaluation/decision support. The tools are basically spreadsheet forms, and are not interactive with the user. It is apparent that considerable thought and effort went into the development of these tools. Unfortunately, the primary goal of the INSIDE Project has yet to be achieved. The application of the INSET toolkit by industry so far has been limited primarily to the sponsoring organizations and companies. This is in part due to the acquisition cost required by the sponsors, which has tempered enthusiasm. However, they could serve as an effective way to jump start a tools development effort by an individual company or an industry group.

The advantage of integrated tools is that the needed EHS collaboration and conflict resolution can be built in before MERITT practitioners use them. In other words, many of the principal EHS strategies are already incorporated in the tool. This reduces resource requirements for implementing MERITT, since the need to assemble large teams to identify and resolve issues is greatly diminished.

## 5.3. Need for Integrated Tools

### 5.3.1. Role of MERITT Tools

One of the crucial needs is having tools that focus on the potential conflicts among EHS disciplines that invariably occur if they are not addressed in an integrated fashion. The importance of this is illustrated in the following case history [4].

In 1995, a fire and explosion involving three tanks of a crude solvent occurred at a chemical terminal facility. The cause of the explosion was attributed to drums of activated carbon connected to the solvent tanks, as required by EPA to prevent environmental emissions. (Subsequently, in May 1997 EPA issued a safety alert regarding use of activated carbon systems with certain types of substances. [4]) The fire and explosion damaged other tanks, generated toxic gases, and resulted in large-scale evacuation of local residents. The message in this is that had the MERITT approach been used with the application of suitable tools, this safety problem most probably would have been identified and eliminated from the pollution control design.

An important related issue with MERITT tools is that there is no common basis for decision making [5]. The results of applying evaluation tools is generally a relative ranking within the specific EHS discipline (e.g., IS). For selection of the best MERITT design, it is desirable to factor the results of applying each index or metric into the decision-making process. This has been approached by applying some type of weighting mechanism so as to consider all the potentially competing interests. The methodology, based on decision analysis, is one that has been applied by some companies [5].

One AIChE guidelines book [24] presents formal decision-making tools that may be useful in multiattribute option selection. These include:

- Voting methods
- Weight scoring methods
- Cost benefit analysis
- Mathematical programming
- Payoff matrix analysis
- Decision analysis
- Multiattribute utility analysis

The latter is particularly useful if the utility functions can be demonstrated with data to establish their validity.

Another tools development issue is achieving flexibility. Tools developed for one industry (e.g., pharmaceuticals) may be less useful for another industry (e.g., petroleum refining). This issue was also identified by the INSIDE project [2]:

> It was recognized that different sectors of industry have very different processes, time-scales, and priorities, and that a flexible set of tools would be needed to meet these very different needs. The tools were aimed at the designers and chemists who are generally best placed to seek out and implement inherent SHE alternatives.

Clearly, there are two possible solutions to this issue. The approach taken by INSIDE was to develop tools that were broadly applicable, and therefore could be applied to a wide range of industrial situations. Alternatively, tools that are more narrowly focused and industry specific, such as Solvent Selection Guidelines, may prove highly useful. Unquestionably, there is value in both types of MERITT tools.

## 5.3.2. Illustrative MERITT Tools

While there currently are few truly integrated MERITT tools, some nonintegrated tools can be easily modified to achieve MERITT objectives. Some possible adjustments to existing tools are illustrated in this section.

### Guide Word MERITT Tool

The use of guide words to initiate creative thinking during a structure brainstorming is commonplace for safety inquiry (e.g., HAZOP studies) and has been promoted for pollution prevention [3]. At least one enterprise [14] is using it for the concept initiation stage for chemical reaction hazard identification. An analogous technique is What-If analysis, which utilizes established what-if questions to identify issues, mostly of a safety nature. By combining these techniques, it is possible to construct a set of MERITT inquiry questions to apply to various EHS characteristics of process design. Table 5-4 illustrates one format for such a tool. The construct consists of a *process attribute* that has an associated *action* or *property* that can be *modified* to potentially enhance the EHS acceptability. The intent is to stimulate thinking about how the process attributes could be modified to achieve the goals of MERITT. The assembled guide words are extracted from the principal strategies of IS, P2, and GC.

This example is provided with the expectation that development of other forms of integrated guide word tools will be sparked by this illustration.

### Total Cost Assessment for MERITT

*Overview*

Total cost assessment (TCA) methodology is a relatively new tool that was developed under the sponsorship of the Center for Waste Reduction Technologies (CWRT). The stated objective of the TCA methodology is to provide a disciplined and standardized approach to improve business decisions by better evaluating the complete realm of potential environmental and health costs that are experienced by enterprises, and that may impact the environment and society [21]. The methodology allows for the evaluation of various categories of cost, both direct (e.g., treat-

*TABLE 5-4.  MERITT Inquiry Tool*

What if the *[affected process attribute]* *[action/property]* is *[modifying guide word]* ?

| Affected Process Attribute | Action/Property | Modifying Guide Word |
|---|---|---|
| Chemistry | Route | Atom economical, low derivative, low waste, less hazardous, inherently safer |
| Product or Reactant | Characteristic | Non/less-toxic, biodegradable, non/less persistent, used in reduced quantity, utilized for process advantage |
| Raw Material | Utilized | Renewable, non/less depleting, non/less toxic, non/less flammable, readily available, reduced |
| Catalyst | Utilized | Highly selective, nonstoichiometric, reduced |
| Solvent | Selected | Eliminated, a reactant, water, non/less flammable, non/less toxic, low volatility, reduced |
| Energy | Consumption | Less, reused, integrated |
| Waste | Quantity/constituent | Eliminated, reduced, reused, recovered, recycled, contained, segregated, managed differently |
| Process | Condition (T, P, conc., corrosivity, etc.) | Non/less severe, altered |
| Process | Equipment | Eliminated, smaller, more intensive, more efficient, simpler |
| Operation (cleaning, use, release) | Frequency/sequence | More/less than customary, changed, automated |
| Storage | Inventory or condition | Eliminated, less, attenuated |

ment) and indirect (e.g., loss of market share), associated with different product or process implementation alternatives. In the case of direct costs, these are routinely allocated in bundles across the enterprise's operating groups, are generally well understood, and are taken into account in project execution. However, many indirect costs relate to future contingent costs, and are frequently neglected or underestimated, if addressed by the project at all. The TCA methodology prompts the practitioner to consider all potential EHS costs. A more complete description of TCA is provided in Appendix A.

The centerpiece of the TCA methodology is the development of a Total Cost Inventory (TCI) that considers costs incurred as well as costs avoided or other benefits, as shown in Figure 5-3. Costs are broadly grouped as recurring and nonrecurring (episodic). For each group, five categories (Types I–V) of costs are defined. A brief description of each category is provided in Table 5-5.

While the initial focus of TCA was skewed toward environmental and health-related costs, the methodology is sufficiently generic that acute safety risks can also be handled. One of the main attractions of incorporating safety-related decision costs into TCA is that it addresses the issue of putting all the EHS project considerations on a level playing field. Cost, after all, is the ultimate equalizer. Furthermore, it is well understood by managers and business decision makers. There is no need to "bring management along" on a new decision-making tool or concept. The challenge is to reduce all the potential EHS impacts associ-

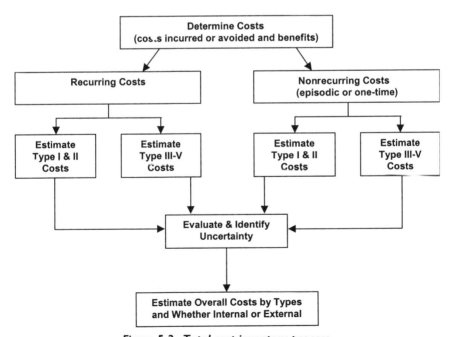

*Figure 5-3. Total cost inventory process.*

**TABLE 5-5. TCA Cost Categories**

| Cost Category | Title | Description |
|---|---|---|
| Type I | Direct Capital and Expense Costs, Nonrecurring | Mainly project-related costs for baseline and alternative designs |
|  | Direct Recurring Manufacturing Site Costs | Operating costs including environmental control and regulatory compliance for baseline and alternative designs |
| Type II | Corporate Overhead Costs | All indirect costs associated with each option |
| Type III | Future and Contingent Liability Costs | Liability costs arising from activities that caused environmental/health impacts to employees or the general public |
| Type IV | Internal Intangible Costs | Cost impact of low staff morale, market share, relationships |
| Type V | External Impact Costs | Cost of environmental cleanup, sustainability metrics, product health impacts |

ated with a particular design option to cost terms. However, TCA guides the practitioner in this area, and has default values for many of the indirect costs, which are less easily estimated. Another advantage is that TCA was designed to integrate with life-cycle analysis (LCA) and life-cycle inventory (LCI), which are generally understood and embraced by many enterprises.

*MERITT Adjustment to TCA*

As mentioned above, the TCA methodology is sufficiently generic that acute safety risks can be handled fairly easily. The main adjustment would be in the range of costs addressed in the Type I–V cost categories. In fact, the cost tables may need only minor adjustments to make some items more explicit. For example, the input table for Type III costs addresses chronic health impact costs for employees and the community. The table can easily be modified to include the cost impact of acute episodic events such as fire, explosion, and toxic vapor release. Similarly, Type IV costs include the license to operate. The cost of business interruption due to an episodic event could be included as well (the TCA algorithms adjust future nonrecurring cost for probability of occurrence).

Some comparisons are possible just by inclusion of the appropriate costs and benefits within existing cost accounts. For example, maintenance related costs are included in Type II costs. In the application of TCA for inherently safer alternatives, the user should take credit for not having to maintain high integrity instrumented safeguarding systems and other engineered systems that would be needed for the less inherently safe option.

### Matrix Assessment for MERITT

Another technique that has been utilized to review designs for MERITT concepts is a numerical ranking matrix [7]. The ranking system can be designed to reflect corporate policies and to focus design teams on broad objectives such as product and workplace safety, environmental protection, and resource conservation. The approach is integrated with LCA by using the same life-cycle stages. For example, Figure 5-4 shows the life-cycle stages defined by an original equipment manufacturer (OEM). The premanufacturing stage is roughly equivalent to the process chemistry stage defined in Chapter 3. Note that the three main EHS tenets form the columns of the matrix. The corporate goals in each area are also highlighted. The scores from each EHS objective are combined into an overall score at each stage for each alternative.

A sample rating sheet for the premanufacturing stage is shown in Figure 5-5. Similar sheets are used for the other stages. The rating sheet is composed of a series of established and standardized questions appropriate for the given life-cycle stage. One of the features of the rating system is that the questions are designed to be answered *yes* or *no*, with a *yes* indicating a potential issue or hazard. This feature addresses one of the weaknesses of rank ordering tools; namely, they are often subjective. This weakness is not completely eliminated, because users still may need a means to calibrate "how much?" justifies a *yes* response. This can lead to the development of internal metrics for some questions to obtain consistency in the application of the matrix ranking system.

| PRODUCT ASSESSMENT MATRIX | | | | | |
|---|---|---|---|---|---|
| **YEAR 2007 GOALS**<br><br>**PRODUCT LIFE-CYCLE STAGES** | 1. Health and Safety<br><br>Zero occupational fatalities<br>Zero serious injuries<br>Reduce lost workday incidence rate to 0.14 | 2. Pollution Prevention<br><br>Reduce Waste:<br>-Not recycled by 60%<br>-Recycled by 35%<br>Reduce air emissions by 60% | 3. Material Conservation<br><br>Reduce energy use by 25%<br>Reduce water use by 25% | WEIGHTS | Weighted Score |
| 1. Premanufacture<br>  Material Selection<br>  Supplier Management<br>  Delivery of Purchased Materials<br>  Packaging<br>  Storage | Score<br>0.00 | Score<br>0.00 | Score<br>0.00 | 0.20 | 0.00 |
| 2. Product Manufacture<br>  Machining/ Forming<br>  Material Processing<br>  Finishing/Cleaning<br>  Surface Treating<br>  Assembling<br>  Testing | Score<br>0.00 | Score<br>0.00 | Score<br>0.00 | 0.20 | 0.00 |
| 3. Distribution & Installation<br>  Packaging<br>  Shipping<br>  Field Testing & Construction<br>  Warehousing | Score<br>0.00 | Score<br>0.00 | Score<br>0.00 | 0.20 | 0.00 |
| 4. Product Use & Service<br>  Normal Use/ Reasonable Misuse<br>  Routine Maintenance<br>  Overhaul & Repair | Score<br>0.00 | Score<br>0.00 | Score<br>0.00 | 0.20 | 0.00 |
| 5. End-of Life<br>  Collection<br>  Reuse/ Remanufacture<br>  Recycle<br>  Disposal | Score<br>0.00 | Score<br>0.00 | Score<br>0.00 | 0.20 | 0.00 |
| **WEIGHTS** | 0.33 | 0.33 | 0.33 | | 0.00 |
| **Weighted Score** | 0.00 | 0.00 | 0.00 | 0.00% | |

*Figure 5-4. Overall rankings.*

| Premanufacturing | Weight | Yes | No | ?? | Comments |
|---|---|---|---|---|---|
| **Safety** | | | | | |
| Do product specifications require the supplier to use an inherently hazardous process? | 15% | | | | |
| Do suppliers ship materials or components that require special equipment or precautions during transport to avoid public injury or exposures? | 15% | | | | |
| Does the product incorporate any toxic or hazardous materials that require special equipment or work procedures to assure safe handling during transport or storage? | 20% | | | | |
| Is any special equipment, such as lifting hooks, fixtures,etc., required to handle delivered materials iin a safe manner? | 15% | | | | |
| Are chemicals and/or other materials ordered and stored in a container size or total quantity that require special precautions? | 15% | | | | |
| Does the product incorporate any of the UTC priority chemicals or other material identifed on a customer black or grey list? | 20% | | | | |
| Total | 0 | | | | |
| **Pollution** | | | | | |
| Do product specifications require the supplier to use any of the listed EH&S sensitive manufacturing processes? | 30% | | | | |
| Do any purchased materials require special handling or equipment during transport to avoid risk of spills or venting? | 10% | | | | |
| Do purchased materials use an excessive amount of packaging that must be managed by the receiving facilities? | 10% | | | | |
| Are chemicals and/or other materials ordered and stored in such a manner that increases risk of spill? | 15% | | | | |
| Do the product specifications call out use of any material, component or process that inherently generates large amounts of waste- either at the supplier's facility or in subsequent manufacturing steps? | 15% | | | | |
| Do any purchased materials or components require special disposal procedures? | 20% | | | | |
| Total | 0 | | | | |
| **Conservation** | | | | | |
| Do product specifications require suppliers to use a process that is energy intensive or requires large volumes of water? | 15% | | | | |
| Does the supply chain impose high transportation costs or rely on insecure sources? | 10% | | | | |
| Do any purchased materials or components require special storage (e.g. refrigeration) that is enrgy intensive? | 15% | | | | |
| Do product specifications restrict the supplier's ability to use recycled or remanufactured content? | 20% | | | | |
| Do any purchased materials or components contain additives, such as bromated flame retardants, or specify a material composition or purity that complicates recycling or limits the use of recycled content? | 20% | | | | |
| Does the product design require field maintenance or overhaul to maintain safe operation and achieve design life? | 20% | | | | |
| Total | 0 | | | | |

*Figure 5-5. Premanufacturing review.*

## 5.4 Development Needs

*The true rule, in determining to embrace, or reject anything, is not whether it have any evil in it; but whether it have more of evil, than good. There are few things wholly evil, or wholly good. Almost everything ... is an inseparable compound of the two; so that our best judgement of the preponderance between them is continually demanded.*—A. Lincoln

Better tools can lead to better decisions. More integrated MERITT tools (viz., Solvent Selection Guide) are needed for various segments of the chemical and other industries. The INSET toolkit has many such tools, but access barriers are limiting its use. Developing public domain tools can help advance a wider adoption of MERITT. Mechanisms that would be desirable in this regard include:

- Development and publishing of company developed MERITT tools and metrics
- Sponsorship of MERITT tool and metric development by professional societies and industry trade associations
- Population of a web site with noncompetitive tools such as generic solvent selection guides, energetics of various chemistries, etc.

Readers of this book are not only invited, but also encouraged, to develop and submit usable tools not included in this text. Areas of particular need include:

- Selection guide for generic chemistries (particularly useful to pharmaceutical and specialty chemical industries)
- Reaction selection guide (a stand-alone tool covering common reactions to assist chemists in choosing the best alternative, in a similar manner to the Solvent Selection Guide)
- MERITT option generation tools (useful to all chemical industry)
- More integrated MERITT tools in all three categories (inquiry, option generation, and option evaluation/decision support)

## 5.5. References

1. Mansfield, D., *INSET Toolkit Stages III and IV—Process Front End and Detailed Design*, London PRP, June 1997.

2. Mansfield, D., The *INSIDE* Project—Inherent SHE in Design, London, June 1997.

3. Curzons, A., Constable, D.C., and Cunningham, V.L., "Solvent Selection Guide: A Guide to the Integration of Environmental, Health, and Safety Criteria into Solvent Selection," in *Clean Products & Processes,* 1997.

4. Mannan, M.S., & Baldwin, J.T., *Inherently Safer Is Inherently Cleaner: A Comprehensive Design Approach*, Chemical Process Safety Report, March 2000, p. 125.

5. Hendershot, D.C., *Measuring Inherent Safety, Health, Environmental Characteristics Early in Process Development*, Process Safety Progress, Summer 1997, pp. 78–79.

6. *The Mond Index, 2nd Ed.,* Imperial Chemical Industries, Winnington, Northwick, Cheshire, U.K., 1985.

7. Swarr, T.S., Legarth, J.B., Huang, E.A., *Implementation of Design for Environment at a Diversified OEM*, AIChE Spring Meeting, Atlanta, 2000.

8. Pojasek, R.B., *Identifying P2 Alternatives with Brainstorming and Brainwriting*, Pollution Prevention Review, Autumn 1996, 6(4): pp. 93–97.

9. *Guidelines for Hazard Evaluation Procedures*, AIChE Center for Chemical Process Safety, 1992.

10. Burk, A.F., *What-If/Checklist—A Powerful Process Hazard Review Technique*, presented at the AIChE Summer National Meeting, Pittsburgh, August 1991.

11. *Inherently Safer Chemical Processes—A Life Cycle Approach,* Concept Book, AIChE Center for Chemical Process Safety, 1996.

12. Mulholland, K.L., and Dyer, J.A., *Pollution Prevention Methodology, Technologies, and Practices,* AIChE Press, 1999.

13. *Guidelines for Design Solutions for Process Equipment Failures,* AIChE Center for Chemical Process Safety, 1998.

14. Mosley, D.W., et. al., *Tools for Understanding Reactive Chemical Hazards Early in Process Development*, presented at AIChE 2000 Spring National Meeting, Atlanta, March 2000.

15. Tyler, B.J., Thomas, A.R., Doran, P., and Greig, T.R., "A Toxicity Hazard Index," *Chemical Health and Safety,* January/February 1996, pp.19–25.

16. Vilchez, J.A., and Casals, J., "Hazard Index for Runaway Reactions," *Journal of Loss Prevention in the Process Industries,* Vol. 4, pp. 125–127 (1991).

17. Cave, S.R., and Edwards, D.W., "A Methodology for Chemical Process Route Selection Based on Assessment of Inherent Environmental Hazard," *The 1997 IChemE Jubilee Research Event,* pp. 49–52.

18. Edwards, D.W., and Lawrence, D., "Assessing the Inherent Safety of Chemical Process Routes: Is There a Relation Between Plant Costs and Inherent Safety?" *Trans. IChemE,* 71B, pp. 252–258 (1993).

19. Edwards, D.W., Lawrence, D., and Rushton, A.G., "Quantifying the Inherent Safety of Chemical Process Routes," 5th World Congress of Chemical Engineering, July 14–18, 1996, San Diego, CA, Vol. II, pp. 1113–1118. New York: American Institute of Chemical Engineers.

20. *Dow's Fire & Explosion Index Hazard Classification Guide*, CEP Technical Manual, AIChE.

21. *Total Cost Assessment Methodology Manual,* AIChE Center for Waste Reduction Technology, June 1999.

22. *Layer of Protection Analysis: Simplified Process Risk Assessment,* AIChE Center for Chemical Process Safety, 2001.

23. *Management of Hazards Associated with Location of Process Plant Buildings*, API Recommended Practice 752, 1st ed., May 1995.

24. *Tools for Making Acute Risk Decisions with Chemical Process Safety Applications,* AIChE Center for Chemical Process Safety, 1995.

25. Halwagi, M. M., *Pollution Prevention through Process Integration*, Academic Press, San Diego, 1997.

26. Mosley, D.W., Ness, A.I., and Hendershot, D.C., "Screen Reactive Chemical Hazards Early in Process Development," *Chemical Engineering Progress,* pp. 51–64, November 2000.

27. Cohan, D., and Gess, D., *Managing Life Cycle Costs,* Decision Focus Incorporated (DFI), Mountain View, CA (1993).

# 6

# Application of MERITT

This chapter presents a worked example to illustrate the application of MERITT and to demonstrate the business advantage of the integration of EHS perspectives that MERITT provides. In the example, the situation that offers the opportunity for employing MERITT is discussed first. Next, the business-as-usual approach to incorporating EHS considerations is described. This is followed by a discussion of the use of MERITT for addressing project EHS characteristics and design issues. A comparison of the approaches that shows the additional business value achieved by using MERITT is then given. A second example that addresses development of new chemistry for a replacement solvent is also discussed.

## 6.1. Introduction

In this chapter, the use of MERITT is demonstrated using a worked example, based on an actual case study. The example concerns a process upgrade and shows how earlier stage process development thinking was incorporated into what is generally a more limited scope project. During the execution of the actual project, the EHS issues were addressed in a nonintegrated fashion. Subsequently, MERITT has been applied to illustrate the advantage of synergy in addressing EHS characteristics and design issues in process development. The use of MERITT results in the identification of additional significant business value.

In this example, the project began in the process definition stage of an upgrading project. Through the application of MERITT, the project conceptually transitioned back several stages to process chemistry and process development, before proceeding into basic process engineering (see Figure 6-1). In actuality, some of these stages overlapped or proceeded in parallel in order to obtain an acceptable process implementation schedule.

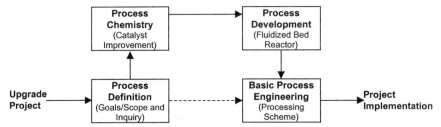

*Figure 6-1. Chemical intermediate upgrade project—stage iteration with MERITT infusion.*

## 6.2. Situation

A chemical intermediate is manufactured by a 30-year-old process that generates an unacceptable level of waste compared to current standards of environmental performance. The facility is located within city limits, and new local environmental regulations require that an expensive end-of-pipe treatment device be installed on a gaseous emission source unless more economical source reduction measures can be identified and implemented. The local community is unaware that the process emits noxious compounds—thus, faced with increased public scrutiny due to new regulations, the business has elected to relocate the manufacturing process to a new site.

The separation steps involved in the existing process include:

1. Aqueous scrubbing of a reactor off-gas to condense and capture the reactant and product,
2. Extraction of the reactant and product from the aqueous phase into benzene solvent, and
3. High-purity distillation of the reactant and product.

Benzene is recovered for recycle, and the aqueous raffinate stream from the extraction step is stripped of benzene before discharge to a biological wastewater treatment plant. The process flow diagram and mass balance for the existing process are shown in Figure 6-2 and Table 6-1, respectively.

### 6.2.1. Waste

The two process waste streams are the gas emissions from the acid scrubber, stream A-10, and the water stream from the benzene extrac-

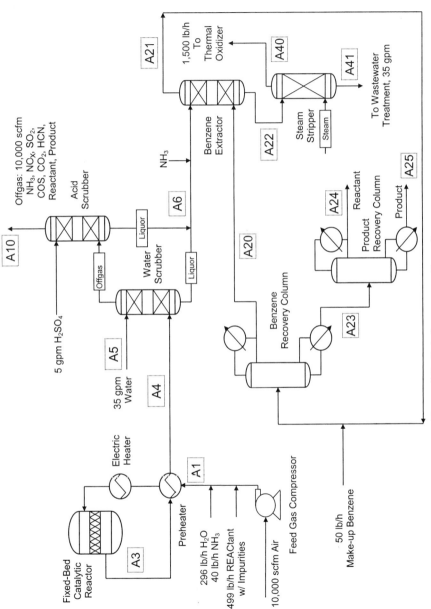

Figure 6-2. Existing process flow diagram.

113

**TABLE 6-1. Existing Process Mass Balance**

| Stream Number | A1 | A3 | A4 | A5 | A6 | A10 | A20 | A21 | A22 | A23 | A24 | A25 | A40 | A41 |
|---|---|---|---|---|---|---|---|---|---|---|---|---|---|---|
| Phase | Vapor | Vapor | Vapor | Liquid | Liquid | Vapor | Liquid | Liquid | Liquid | Liquid | Liquid | Liquid | Vapor | Liquid |
| Component | | | | | | Mass Flow (lb/hr) | | | | | | | | |
| Benzene ($C_6H_6$) | 0 | 0 | 0 | 0 | 0 | 0 | 4949.6 | 4949.6 | 33.4 | 0.0089 | 0.0089 | 0 | 33.4 | 0 |
| Nitrogen ($N_2$) | 36952 | 36952 | 36952 | 0 | 0.22 | 36952 | 0.22 | 0.22 | 0.0024 | 0 | 0 | 0 | 0.0024 | 0 |
| Oxygen ($O_2$) | 11298 | 11064 | 11064 | 0 | 0.053 | 11064 | 0.053 | 0.053 | 0.00018 | 0 | 0 | 0 | 0.00018 | 0 |
| Water ($H_2O$) | 296 | 433.3 | 433.3 | 17500 | 14821 | 3112.2 | 15.9 | 15.9 | 14820 | 0 | 0 | 0 | 1500 | 16331 |
| REACtant | 483 | 205.8 | 205.8 | 0 | 202.8 | 3 | 2.7 | 203.1 | 1.7 | 200.4 | 199.3 | 1 | 1.7 | 0.0009 |
| PRODuct | 1 | 207.7 | 207.7 | 0 | 207.7 | 0 | 1.4E-05 | 206.6 | 1.1 | 206.6 | 1 | 205.6 | 0.054 | 1 |
| IMPURity | 15 | 15 | 15 | 0 | 2.4 | 12.6 | 0.52 | 2.4 | 0.013 | 1.9 | 1.9 | 0.0016 | 0.013 | 0 |
| Ammonia ($NH_3$) | 40 | 8 | 8 | 0 | 0 | 0 | 0.13 | 0.13 | 7.9 | 0 | 0 | 0 | 7.8 | 0.03 |
| $NO_x$ | 0 | 14 | 14 | 0 | 2.3 | 11.7 | 0.21 | 0.21 | 2.1 | 0 | 0 | 0 | 2.1 | 0.004 |
| Sulfur Dioxide ($SO_2$) | 0 | 39.1 | 39.1 | 0 | 2.9 | 36.2 | 0.27 | 0.27 | 2.6 | 0 | 0 | 0 | 2.6 | 0.002 |
| Carbonyl Sulfide (COS) | 0 | 18.3 | 18.3 | 0 | 0.0018 | 18.3 | 1.6E-04 | 1.6E-04 | 0.0016 | 0 | 0 | 0 | 0.0016 | 0 |
| Carbon Dioxide ($CO_2$) | 0 | 120.9 | 120.9 | 0 | 0.025 | 120.9 | 0.0023 | 0.0023 | 0.022 | 0 | 0 | 0 | 0.022 | 0 |
| Hydrogen Cyanide (HCN) | 0 | 16.5 | 16.5 | 0 | 4.3 | 12.2 | 0.4 | 0.4 | 3.9 | 0 | 0 | 0 | 3.5 | 0.34 |
| Sulfuric Acid ($H_2SO_4$) | 0 | 0 | 0 | 0 | 26 | 0 | 0 | 0 | 0 | 0 | 0 | 0 | 0 | 0 |
| Ammonium Sulfate (($NH_4$)$_2SO_4$) | 0 | 0 | 0 | 0 | 30 | 0 | 0 | 0 | 65 | 0 | 0 | 0 | 0 | 65 |
| **Total** | 49085 | 49094.6 | 49094.6 | 17500 | 15299.7 | 51343.1 | 4970.00 | 5378.89 | 14937.7 | 408.909 | 202.209 | 206.602 | 1551.19 | 16397.4 |
| Temperature (°C) | 266.5 | 400 | 352 | 30 | 54.2 | 48.6 | -65 | 53.9 | 49.7 | 139.6 | 31.7 | 136.3 | 105.4 | 108.9 |
| Pressure (lb/in² absolute) | 75 | 31 | 29 | 25 | 18 | 18 | 9.5 | 14.7 | 14.7 | 10 | 0.5 | 0.65 | 18 | 20 |
| Molecular Weight (g/g-mole) | 28.9 | 28.9 | 28.9 | 18 | 18.4 | 27.9 | 77.3 | 78.8 | 18.1 | 104.3 | 99 | 109.9 | 18.4 | 18 |
| Density (lb/ft³) | 0.21 | 0.069 | 0.069 | 62.9 | 61.8 | 0.081 | 60.2 | 53.7 | 61.9 | 64.9 | 66.1 | 69.7 | 0.046 | 58.9 |

tor, stream A-22. The major characteristics of the two waste streams are:

- Stream A-10—high volumetric flow as a result of using air as a source of oxygen for the reaction and the need for nitrogen as a heat sink to absorb the heat of reaction and to prevent hot spots in the ceramic catalyst. The contaminants in the gas stream are byproducts of reaction and excess reactants such as ammonia.
- Stream A-22—high volumetric flow due to using water as a coolant and scrubbing agent. The contaminants are byproducts of reaction and benzene, the extracting agent.

## 6.2.2. Safety Hazards

The three major safety hazards involve the reactor system and the use of benzene as the extracting agent. The safety hazards are:

- *Reactor*—Operates above ambient pressure and temperature conditions and involves an exothermic reaction. The high volumetric flow and the presence of toxic compounds, such as hydrogen cyanide and sulfur dioxide, further increase the hazard.
- *Benzene*—Solvent material is carcinogenic and very flammable. Extra care is required in storage and transport not only to prevent flammable conditions, but also to prevent fugitive emissions.
- *Scrubber exit vapor*—The high volumetric flow will require a large abatement device such as a thermal oxidizer, which has associated safety hazards. In addition, the untreated stream contains toxic compounds.

## 6.2.3. Green Chemistry Issues

From a green chemistry perspective there are two major concerns with the reactor:

- *Toxicity*—the formation of toxic byproducts in the reactor.
- *Complex Solvent Use*—not only is benzene toxic, but its use is associated with many additional EHS burdens.

## 6.3. Nonintegrated EHS Approach

This section presents a plausible outcome to the solution of the environmental problems described above, had the traditional approach to the infusion of EHS practices been followed. In this case, the primary driver was environmental pollution; hence, P2 practitioners would have been directly involved. At some point (albeit rather late in the process) it is assumed that a process safety engineer would have been asked to review the design, either formally (via a process hazards analysis) or informally, as a review team member. It is further assumed that the safety engineer and a process chemist would not have been included in the brainstorming of initial ideas to improve the environmental footprint.

### 6.3.1. Set Goals

The team first set its goals, which helped analyze the drivers for pollution prevention and develop the appropriate criteria to screen the options that would be generated during the subsequent brainstorming session. The goals for the case study were to: reduce the nitrogen content in the plant outfall; minimize air toxic emissions; and lower the capital investment and cost of manufacture for the new facility. Note that the goals were primarily limited to P2 issues.

### 6.3.2. Define the Problem

The team then defined the problem in order to understand the targeted waste streams and the manufacturing process steps that generate them. Two techniques were used for examining the waste generation problem: *waste stream analysis* and *process analysis* which are both discussed in Chapter 3 of Mulholland and Dyer's book "Pollution Prevention: Methodology, Technologies, and Practices"[1] and are not revisited here. The goal of both techniques is to frame the problem such that pertinent questions arise. When the right questions are asked, the more feasible and practical solutions for pollution prevention become obvious.

### 6.3.3. Identify Options

The core project team, supported by pollution prevention specialists, collaborated to identify ideas to achieve the defined goals. A systematic

**TABLE 6-2.** *Selected Initial Ideas*

---

### Acid Scrubber Off-Gas Stream (A10)

- Use enriched air or pure oxygen in the reactor instead of air to reduce or eliminate $N_2$.
- Recycle the acid scrubber off-gas (stream A10) to the reactor and add pure $O_2$ as make-up.
- Use catalyst bed intercooling to limit temperature rise that leads to unwanted byproduct formation.
- Add a second heat exchanger in series with the preheater to improve energy recovery.
- Use a different heat sink instead of nitrogen, such as $CO_2$ or steam.

### Wastewater Stream from Benzene Extractor (A22)

- Combine the acid and water scrubbers to reduce water consumption.
- Recycle the water stream from the benzene extractor (A22) to the water scrubber.
- Recycle the water stream from the steam stripper (A41) to the water scrubber.
- Use caustic soda instead of ammonia for pH control of the water stream to the benzene extractor (stream A6) to reduce total nitrogen load to wastewater treatment.
- Replace benzene with a different extraction solvent, such as toluene or xylene.
- Freeze-crystallize REAC and PROD from the water to improve product recovery and reduce total nitrogen load to wastewater treatment.
- Use a multieffect evaporator to concentrate stream A41.
- Use the chilled reactant (REAC) to scrub itself from the reactor off-gas.
- Use a condenser and decanter to separate and recycle the benzene from the steam stripper overheads instead of a thermal oxidizer.

---

methodology, much like the one described by Mulholland and Dyer [1], that uses a structured brainstorming process (requiring minimum resources) was applied to identify possible chemistry and engineering changes to the process. Many possible options were generated and a list of the more feasible ones is provided in Table 6-2.

## 6.3.4. Screening of Options

A first-cut screening of the ideas was performed during the brainstorming session to take advantage of the experience and "gut feel" of the participants in the room. No idea is ever eliminated completely, but it is important to make an initial ranking to help the core assessment team

Some methods that have been used successfully to screen options include:
- Judge yes or no,
- Rank high, medium, or low against specific criteria, and
- Multivoting, where each team member is allotted ten points that can be allocated at their choosing.

focus on those ideas with the highest likelihood of success. The core team, however, was responsible for reconsidering all ideas in a separate session. Because the brainstorming session participants are normally weary by the time the first-cut screening is done, simple methods are used to rank the ideas. Criteria mentioned above in the "Set Goals" section were used for ranking options for the case study.

During the first-cut screening, the number of ideas was reduced by about half, then in a subsequent meeting, the core assessment team revisited the options generated during the brainstorming process. A second-cut screening was made to further narrow the number of "feasible" options that were carried forward and evaluated in further detail. (The collective experience of several professionals shows that about 10–15% of the original ideas make it through a second-cut screening.)

In this case study, once the idea generation phase was complete, the ideas were reexamined and ranked using a yes/no rating method and the following criteria:

- Technical feasibility,
- Economic viability, and
- Waste reduction potential.

Table 6-3 briefly summarizes the process improvement ideas that survived the first-cut screening. During the screening process, a safety engineer was involved to consider inherent safer aspects. One modification was to replace air with enriched air from a pressure swing adsorption unit or oxygen to reduce the nitrogen rate.

## 6.3.5. Idea Evaluation

These ideas were subjected to more detailed evaluations and eventually three options were selected as technically feasible and meeting most of the defined goals. The EHS and estimated cost characteristics of the selected options are summarized in Table 6-4.

**TABLE 6-3. First Cut Screening Ideas**

| | |
|---|---|
| **Recycle Vent Stream and Add Additional O$_2$ as Needed** | |
| Cost | $200,000 investment, $500,000/yr operating cost |
| Benefit | Reduced end-of-pipe investment of $700,000<br>Reduced operating cost of $70,000/yr |
| Waste Minimization | Reduced gas flow to be treated |
| Energy Conservation | Reduced electricity requirements |
| Probability of Success | 90% |

| | |
|---|---|
| **Recycle Benzene from the Steam Stripper Overhead Stream** | |
| Cost | $100,000 investment, $10,000/yr operating cost |
| Benefit | Reduced operating cost of $90,000/yr |
| Waste Minimization | Reduced gas rate to thermal oxidizer |
| Energy Conservation | No change |
| Probability of Success | 90% |

| | |
|---|---|
| **Use Steam Stripper Bottoms as the Source of Water for the Water Scrubber** | |
| Cost | $300,000 investment, $10,000/yr operating cost |
| Benefit | Water conservation and reduced treatment cost |
| Waste Minimization | Reduced wastewater treatment load (35 gpm to 1 gpm) |
| Energy Conservation | No change |
| Probability of Success | 90% |

| | |
|---|---|
| **Use Superheated Steam (as the inert gas) and Pure O$_2$ Instead of Air** | |
| Cost | $2,000,000–$6,700,000 investment,<br>$800,000–$3,500,000/yr operating cost |
| Benefit | Reduced end-of-pipe investment of $800,000<br>Reduced operating cost of $300,000/yr |
| Waste Minimization | Reduced gas flow to be treated |
| Energy Conservation | Reduced electricity and steam requirements |
| Probability of Success | 30% |

| | |
|---|---|
| **Replace the Water Scrubber with a Solvent Scrubber** | |
| Cost | $1,300,000 investment, $100,000/yr operating cost |
| Benefit | Reduced operating cost of $700,000/yr |
| Waste Minimization | Reduced wastewater treatment load (35 gpm to 1 gpm) |
| Energy Conservation | Reduced steam requirements |
| Probability of Success | 90% |

*Table continues on page 120*

**TABLE 6-3.** *First Cut Screening Ideas (continued)*

---

### Use Freeze Crystallization to Separate REACtant, PRODuct, and Water

| | |
|---|---|
| **Cost** | $3,500,000 investment, $1,500,000/yr operating cost |
| **Benefit** | Reduced operating cost of $700,000/yr |
| **Waste Minimization** | No change |
| **Energy Conservation** | No change |
| **Probability of Success** | 40% |

---

### Purify REACtant and PRODuct Directly from the Water Phase, Rather Than Using Extraction

| | |
|---|---|
| **Cost** | $10,500,000 investment, $4,500,000/yr operating cost |
| **Benefit** | Reduced operating cost of $700,000/yr |
| **Waste Minimization** | Reduced wastewater treatment load (35 gpm to 1 gpm) |
| **Energy Conservation** | No change |
| **Probability of Success** | 70% |

---

### Use Fluidized-Bed Reactor (reduces air volume)

| | |
|---|---|
| **Cost** | $500,000 investment, $500,000/yr operating cost |
| **Benefit** | Reduced end-of-pipe investment of $800,000<br>Reduced operating cost of $40,000/yr |
| **Waste Minimization** | Reduced gas flow to be treated |
| **Energy Conservation** | Reduced electricity requirements |
| **Probability of Success** | 80% |

---

### Use New Catalyst with Better Selectivity/Conversion

| | |
|---|---|
| **Cost** | Unknown |
| **Benefit** | Unknown |
| **Waste Minimization** | Reduces waste generation by 50% for a 5% increase in yield |
| **Energy Conservation** | None |
| **Probability of Success** | 50% |

---

### Change Air-to-Feed Ratio to Reactor to Reduce COS Generation

| | |
|---|---|
| **Cost** | Unknown |
| **Benefit** | Unknown |
| **Waste Minimization** | None |
| **Energy Conservation** | None |
| **Probability of Success** | 50% |

---

### Recover and Recycle Tars from REACtant/PRODuct Splitter

| | |
|---|---|
| **Cost** | Unknown |
| **Benefit** | Unknown |
| **Waste Minimization** | Eliminate hazardous waste incineration |
| **Energy Conservation** | None |
| **Probability of Success** | 70–80% |

### 6.3.6. Results

As Table 6-4 indicates, the waste minimization goal was achieved with a net capital savings of $100,000. However, net operating cost increased by $340,000 per year. While the P2 and capital cost goals were basically achieved, the added operating costs meant that no overall business value was obtained. Management's view might be described as underwhelmed, since additional costs were required for pollution control.

## 6.4. MERITT Approach

This section demonstrates the use of the five step MERITT structure as presented in Chapter 4, Figure 4-3.

### 6.4.1. Establishing the Basis

Because this example involved an existing process, many aspects of the product and process design were well established. Much of the information for establishing the basis is contained in the initial process design documentation and operating plant improvements. From this information, the project team can extract the main objectives, criteria, and constraints as shown in Table 6-5.

*TABLE 6-4. Results of Idea Evaluation*

| Option | Waste Minimization | Net Capex $K | Net Opex $K/yr |
|---|---|---|---|
| Recycle vent stream and add enriched air as needed | Reduced gas flow to be treated by 80% | (500) | 430 |
| Recycle benzene from the steam stripper overhead stream | Reduced gas rate to thermal oxidizer | 100 | (90) |
| Use steam stripper bottoms as a source of water for the water scrubber | Reduced water treatment load from 35 gpm to 1 gpm | 300 | ~0 |
| | Total | (100) | 340 |

*TABLE 6-5. Development Basis*

| Element | Basis |
|---|---|
| **Objectives** | |
| General Process | Production capacity |
| | Product efficacy |
| | Plant availability/reliability |
| | Investment cost |
| MERITT | Produce zero manufacturing waste |
| | Develop molecules that are not persistent, toxic, and bioaccumulative |
| | Nonhazardous manufacturing process (i.e., low toxicity, explosivity, and reactivity) |
| **Criteria** | |
| General Process | Product quality |
| | Operational efficacy |
| | Process economics |
| | High controllability |
| MERITT | Limit water use |
| | Limit hazardous byproducts |
| | Limit toxic solvent use |
| **Requirements/Constraints** | |
| General Process | Corporate design standards |
| | Plant commercialization date |
| | Utilities availability/cost |
| MERITT | Environmental regulations |
| | OSHA regulations |
| | Community acceptance |

A benefit of establishing the basis is the identification of common goals among the EHS practices. The common goals highlighted the need to better coordinate the application of these practices in designing the upgraded process.

## 6.4.2. Identifying Issues

Issue identification is a critical second step of MERITT. As is the case with hazard identification, issues that are not uncovered cannot be addressed. The use of inquiry tools can be effective in identifying issues. The example proceeds with the application of the modified What-If inquiry tool presented in Table 5-4 of Chapter 5. The tool would be applied at the team brainstorming session, prior to proceeding with option generation. The results of the teams findings are presented in Table 6-6.

**TABLE 6-6.** *Key Issues Identified*

| Affected Process Characteristic | Inquiry Guide | Finding |
|---|---|---|
| Raw Material | What if air utilization is reduced? | The nitrogen and excess oxygen act as heat sinks to prevent hot spots in the fixed catalyst bed. |
| Solvent | What if water is eliminated? | Water is a cheap, universal solvent and heat-transfer fluid. The water serves two functions in the existing process—to cool the hot reactor gases and to scrub the condensed reactant and product from the gas stream. When the original process was developed, the impact on wastewater treatment was not taken into consideration. |
| Solvent | What if benzene is eliminated? | Benzene was selected due to its high affinity for the reactant and product. Solvent circulation rate would increase with other potential solvents such as toluene. |
| Product/Reactant | What if product/reactant characteristic is utilized for process advantage? | The product solidifies at 50–60°C. The reactant has a normal boiling point at above 120°C and a freezing point below −20°C. These properties, together with increased concentrations at reduced air flow, would allow for direct condensation of the bulk of the product. |
| Process Equipment | What if catalyst bed mixing is intensified? | Better heat transfer and process temperature control, elimination of catalyst hot spots. |

### 6.4.3. Developing Options

For this project, the tool selected for developing options was brainstorming. When the core assessment team had a good understanding of the process and the characteristics of each EHS issue, it convened to brainstorm for ideas. To generate all of the possible ideas for process improvement and MERITT objectives, the core team needed to involve additional talents and diverse points of view. For the case study, the extended brainstorming team included, in addition to the core team, a separations specialist, a pollution prevention and environmental specialist, a process safety engineer, a reaction engineering specialist, a process chemist, and a business expert.

For this case study, the brainstorming session produced over 100 ideas for process improvement. Some of the ideas are listed in Table 6-7. The key issues presented in Table 6-2 above proved to be the "catalyst" to improve the process. Note that many of the ideas are the same for the business-as-usual approach. However, there are several additional concepts resulting from the synergy of new perspectives on the team. As will be seen, some of the new ideas proved to be "breakthrough" concepts that created real business value.

### 6.4.4. Assessing Options

*Screening of Ideas*

The screening of the ideas can be performed in the same way as described in Section 6.3.4 above. There is nothing inherent in the MERITT approach that would preclude this technique. Alternatively, some of the other EHS tools and metrics discussed in Chapter 5 can be utilized. Criteria mentioned above in the "Establishing Basis" section can be used for ranking options for the case study.

During the first-cut screening, the number of ideas was reduced by about half, then in a subsequent meeting, the core assessment team revisited the options generated during the brainstorming process. A second-cut screening was made to further narrow the number of "feasible" options that were carried forward and evaluated in further detail.

Table 6-8 summarizes additional process improvement ideas that survived the first-cut screening. For each screened idea, the team prepared a brief description of the process, chemistry, or technological

***TABLE 6-7. Options Generated from Brainstorming***

| |
|---|
| **Acid Scrubber Off-Gas Stream (A10)** |
| • Use pure oxygen in the reactor instead of air to eliminate $N_2$. |
| • Recycle the acid scrubber off-gas (stream A10) to the reactor and add pure $O_2$ as make-up. |
| • Use a new or improved catalyst in the reactor. |
| • Replace the existing fixed-bed reactor with a fluidized-bed reactor to eliminate hot spots that lead to unwanted byproduct formation. |
| • Add a second heat exchanger in series with the preheater to improve energy recovery. |
| • Use a different heat sink instead of nitrogen, such as $CO_2$ or steam. |
| • Use an indirect-contact heat exchanger to cool the reactor off-gas and condense the product and reactant. |
| **Wastewater Stream from Benzene Extractor (A22)** |
| • Combine the acid and water scrubbers to reduce water consumption. |
| • Recycle the water stream from the benzene extractor (A22) to the water scrubber. |
| • Recycle the water stream from the steam stripper (A41) to the water scrubber. |
| • Use caustic soda instead of ammonia for pH control of the water stream to the benzene extractor (stream A6) to reduce total nitrogen load to wastewater treatment. |
| • Replace benzene with a different extraction solvent, such as toluene or xylene. |
| • Freeze-crystallize REAC and PROD from the water to improve product recovery and reduce total nitrogen load to wastewater treatment. |
| • Use a multieffect evaporator to concentrate stream A41. |
| • Use the chilled reactant (REAC) to scrub itself from the reactor off-gas. |
| • Use a condenser and decanter to separate and recycle the benzene from the steam stripper overheads instead of a thermal oxidizer. |

changes; concerns; estimated capital investment and/or operating cost; benefits if implemented; resources required; timing; and probability of success.

### Evaluation of Selected Ideas

This step of the MERITT process entails a more detailed and thorough technical and economic evaluation of the selected ideas. The first two levels of screening just described minimize the number of feasible alternatives that need to be evaluated in detail.

In general, revised mass and energy balances, process flow diagrams, and operating requirements are generated for each selected idea.

**TABLE 6-8.  *Additional Ideas from First-Cut Screening***

| **Use Fluidized-Bed Reactor (reduces air volume)** | |
| --- | --- |
| **Cost** | $500,000 investment, $500,000/yr operating cost |
| **Benefit** | Reduced end-of-pipe investment of $800,000 |
| | Reduced operating cost of $40,000/yr |
| **Waste Minimization** | Reduced gas flow to be treated |
| **Energy Conservation** | Reduced electricity requirements |
| **Probability of Success** | 80% |
| **Use New Catalyst With Better Selectivity/Conversion** | |
| **Cost** | Unknown |
| **Benefit** | Unknown |
| **Waste Minimization** | Reduces waste generation by 50% for a 5% increase in yield |
| **Energy Conservation** | None |
| **Probability of Success** | 50% |
| **Install Condenser to Remove Product** | |
| **Cost** | $500,000 investment |
| **Benefit** | Unknown |
| **Waste Minimization** | No change |
| **Energy Conservation** | Reduced separation requirements |
| **Probability of Success** | 90% |

Based on these inputs, the EHS characteristics, the investment, operating cost, and net present value for each option can then be determined and compared against the existing process.

### Reconciling Issues and Decisions

At this point, the results of MERITT are fed back into the process development management structure for review, reconciliation of issues, and decisions regarding forward action. Typically this would occur at a stage gate in the project management structure. Although not applied in this case study, tools such as Life-Cycle Analysis (LCA) or Total Cost Assessment (TCA) could be used at this point, to better sharpen the EHS benefits of each option. Some of the EHS index tools presented in Chapter 5 could also be used at this point, to highlight differences in EHS characteristics.

## 6.4.5. Improved Process

The new process (Figure 6-3) consists of a closed-loop fluid-bed reactor. The feed materials, REACtant, oxygen, and ammonia are introduced at the bottom of the reactor. The reactor exit gas is cooled to remove most of the product. The noncondensable vapor is recycled to the reactor. A purge stream is sent to a chilled spray condenser loop. The chilled spray is primarily the reactant, REAC, with small quantities of the product, water, and other dissolved gasses. The condensed liquid is sent to a distillation column where the product is separated from the other constituents that are sent to a water/REAC separation column. The recovered REACtant material is recycled back to the reactor loop and the water is sent to wastewater treatment.

## 6.4.6. MERITT Achievements

*Practices Applied*

Using MERITT to address the issues described in Section 6.2, the team achieved some remarkable results by combining the practices of the IS/GC/P2 disciplines. True synergy was achieved since it is obvious that the results show an overlapping of the different EHS strategies. For example:

- The concept of using a new catalyst in a fluidized-bed system is the fusion of the following IS and GC strategies:
  —Use more moderate process conditions
  —Avoid runaway reaction potential
  —Reduce toxic byproduct formation and lower raw material waste

- The concept of using condensed reactant to recover the product fuses all the different strategies:
  —Avoid introduction of waste materials (water)
  —Simplify the process (elimination of benzene extraction), including going from 7 vessels and associated pumps to 4
  —Reduced volume in the process
  —Eliminate toxic solvents (benzene)

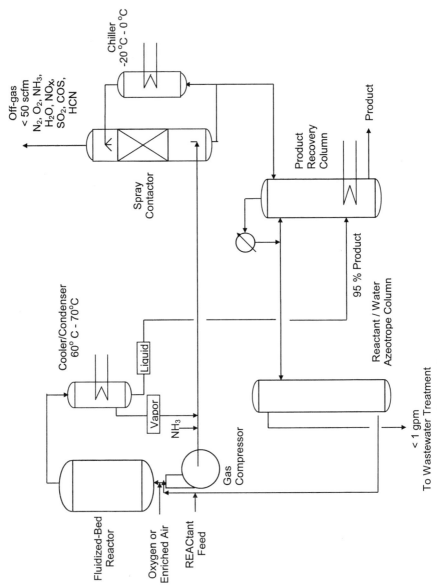

**Figure 6-3.** *Improved process flow diagram.*

Off-gas
< 50 scfm
$N_2$, $O_2$, $NH_3$,
$H_2O$, $NO_x$,
$SO_2$, COS,
HCN

Chiller
-20°C - 0°C

Spray
Contactor

Product
Recovery
Column

Product

Cooler/Condenser
60° C - 70°C

95 % Product

Reactant / Water
Azeotrope Column

Liquid

Vapor

$NH_3$

Gas
Compressor

Fluidized-Bed
Reactor

Oxygen or
Enriched Air

REACtant
Feed

< 1 gpm
To Wastewater Treatment

*Improvements in EHS Characteristics*

The new process addresses each area of concern as follows:

- *Waste reduction*—The condenser in the reactor loop removes 95% of the product, thus minimizing any losses of product as it recycles back into the reactor. The small side stream is cooled and scrubbed by the reactant which eliminates the water scrubbers and the benzene extractor, resulting in a less than 1 gpm wastewater stream. Also, the small side stream results in a small treatment device. The fluidized reactor reduces the gas flow by 70%, and that, combined with maintaining the reactor loop at high pressure, reduces the compressor energy requirements by 85% and thus any waste associated with the making of that energy.
- *Inherently safer process*—The fluidized bed with its lower gas flow results in more uniform temperature, thus reducing a possible runaway reaction. The use of pure oxygen instead of enriched air would have lowered the gas flows more, but at a higher risk and with the expense of extra controls. This compromise was a result of the integration of EHS disciplines brought about by MERITT. The removal of the water scrubber and its attendant solvent extraction system removes a very large fire hazard. Having fewer equipment units further reduces operational hazards.
- *Greener chemistry*—The fluidized bed with its more uniform temperature control should reduce the number of undesirable byproducts resulting from hotspots.

Table 6-9 summarizes the differences with and without MERITT.

*Business Value Creation*

The lower gas processing rates and elimination of the extraction step, benzene handling, and raffinate stripping operations provided significant project investment savings and lower operating costs. When compared to the cost of reproducing the original process, the net investment savings alone were $7.8 million—after including the cost of developing the new design. In addition, the annual operating cost was decreased by $1.5 million. (The total cost was roughly $16 million.) This result generated real excitement for the business managers, and kudos for the MERITT-infused project team.

**TABLE 6-9.** *Differences in Process Design and Operations*

| Independent Reviews | | With MERITT | |
|---|---|---|---|
| Recycle vent stream and add enriched air | • 80% reduction in treated gas flow | More selective catalyst | • 5% increase in yield with 50% decrease in waste generation<br>• Reduced toxic byproduct formation |
| Recycle benzene from steam stripper overhead | • Reduced gas treated in thermal oxidizer | Fluidized-bed reaction system | • Reduced gas flow to be treated<br>• Reduced electricity consumption<br>• Avoided runaway reaction potential |
| Recycle steam stripper bottoms as water source for scrubber | • Reduced wastewater treatment load from 35 to 1 gpm | Add condenser to recover product | • Eliminated benzene extraction step and associated hazards<br>• Reduced energy consumption for separations<br>• Simplified process<br>• Reduced process water requirements and waste generation |
| Reduced capital investment with increase in operating cost | | Significantly reduced capital investment with decrease in operating cost | |

This enhanced business value was created without adversely affecting the project schedule, and for minimal additional resource costs ($25,000).

## 6.5. Product Development Example

The next example illustrates the application of MERITT at the beginning of a product development cycle.

### 6.5.1. Situation

This example involves the synthesis of an orally active HIV-1 protease inhibitor, Merck L-723,524 available as Crixivan®. Since therapeutic doses of this drug are in gram quantities and the molecule contains five chiral centers, manufacturing large quantities has been a huge challenge. The basic synthesis process for the intermediate allyl acetonide involves 16 steps and two water soluble solvents (isopropyl alcohol (IPA) and tetrahydrofuran). The use of tetrahydrofuran (THF) in pharmaceutical processing is problematic as the solvent is used for a range of reactions which must be carried out under anhydrous conditions. However, the desired product is often obtained by aqueous precipitation. Because THF is soluble in water, it typically is recovered by distillation, which produces wastewater and results in loss of antioxidants, such as butylated hydroxytoluene which are added to arrest the potentially dangerous formation of peroxides. The volatility of THF also presents a potential environmental emission source.

## 6.5.2. Use of MERITT

During the concept initiation stage use of a *solvent selection tool* drove researchers to use solvents which were a compromise between process efficacy and environmental acceptance. During the process chemistry stage, the issues of scale-up and large-scale manufacture created opportunities and challenged the development team. By adopting a holistic view of the process and EHS characteristics, a group of researchers and specialists considered the options. Using *inquiry tools,* it became apparent that use of THF and IPA had several disadvantages:

- *Solvent water solubility*—increasing losses through water discharges and energy consumption for solvent recovery
- *Solvent volatility*—increasing losses through vapor emissions
- *Solvent flammability*—increasing fire potential
- *Solvent reactivity*—potential peroxide formation (THF)
- *Solvent toxicity*—moderate health hazard (THF)

Note that the various issues definitely encompass P2 and IS perspectives, but, as will be shown, the final elegant solution involved the infusion of green chemistry.

Once these crucial factors were understood, the subsequent brainstorming focused on reducing or eliminating these disadvantages. Among the suggested alternatives was the use of a "designer solvent" specifically developed to address the MERITT identified issues. The above issues list was converted into a preferred solvent specification, and with the use of resources at the Massachusetts Institute of Technology [2, 3], work was initiated to develop a tailored solvent for the Crixivan process. The result of this research identified n-alkyl tetrahydrofurfural ethers (nATE) as a promising class of solvents, and ultimately n-octal TE (nOTE) as the preferred solvent. Physical property advantages include:

- Negligible water solubility
- Low volatility (boiling point 259°C)

The process advantages of nOTE are significant:

1. *Process simplification*—use of single solvent for the entire synthesis
2. *Decreased energy consumption*—eliminated solvent/water separations
3. *Decreased environmental burden*—eliminated solvent losses to air and water
4. *Increased process safety*—reduced flammability properties, eliminated IPA
5. Reduced health hazards.

Due to items 1 and 2 above, the number of basic process steps required to produce the intermediate allyl acetonide decreased from 16 to 8, and energy consumption was drastically reduced. These process modifications produced real cost benefits in the form of lower initial plant investment and reduced utility and maintenance costs.

Clearly, chemistry and the role of the chemists was paramount in the formulation and execution of the ultimate solution. However, the perspectives of the P2 and IS disciplines acted as a catalyst to spur thinking about how to reduce or eliminate the undesirable properties of the solvent systems being considered. In this case, the final product was clearly greater than the sum of the individual parts.

## 6.6. References

1. Kenneth L. Mulholland and James A. Dyer, *Pollution Prevention Methodology, Technologies, and Practices*, AIChE, 1999.

2. Chin, B., Cermenati, L., Thien, M.P., and Hatton, T. A., "Tailored Solvent for Pollution Prevention in the Pharmaceutical Industries," 3rd Joint China/US Chemical Engineering Conference, Bejing, September 2000.
3. Patent PCT/US96/18237, Replacement Solvents for Use in Chemical Synthesis.

# 7

# Implementation Guidance

For MERITT to be successful, it is critical to overcome cultural and institutional barriers through training, top-down commitment, bottom-up guerilla warfare, reinforcement of desired behaviors, etc. This chapter provides guidance on dealing with a company's formal institutional structure (i.e., its organizational structure, management style, policies, and information systems) and its culture and customs (i.e., the unwritten rules that govern everyday operations and the protection of both the status quo and entitlements—"the project manager is king"). Employing metrics, both integrated and more traditional, to ensure that MERITT is effectively utilized is also discussed.

## 7.1. Incorporating MERITT into an Existing Process Development Process

To ensure MERITT adds value within the context of an existing process development process or system will generally require some modifications and tailoring of each company's existing practices. Such modifications may include:

- MERITT inputs at stage gates and/or milestones by developing a broader set of criteria (see Chapter 4) and/or reviewers, approvers, or even gatekeepers. The inputs will not be the same for each stage gate, but there should be MERITT involvement at each stage gate.
- Identifying minimum expectations for the involvement of EHS expertise and reviews (i.e., applications of tools) at different stages, just as some level of cost estimate is typically required at most stages.
- Adding MERITT-related sign-off's or checklist items for each stage just as they exist for engineering or finance. If your process development process relies more on milestones than stage gates, then

MERITT-related milestones should be identified and integrated into the timeline.

- Providing overview training on MERITT concepts in any formal or informal project manager training. Specific training on MERITT tools and guidance on the means to obtain additional information, whether it is through internal resources or external web sites, is also needed.
- Publicizing a listing or directory of MERITT contacts and resources to assist project managers and other team members.
- Changing your project management metrics and tracking system to also include metrics related to MERITT.
- Adding MERITT to the topics of discussion at routine project meetings. It is vital to ensure that MERITT communications are both routine and expected in order to demonstrate that MERITT is an integral and ongoing part of the product development process.

The most effective means to institute these changes will vary by company, but must reflect each company's culture.

### "Show-Me" Cultures

In many companies it will be important to have a phased roll-out, so that success can be demonstrated internally before widespread adoption occurs. The diverse range of cultures where such an approach works includes companies with:

- An attitude of healthy skepticism
- Hesitancy about trying "new" things
- A "show me that it works here" mindset
- Past experience where a new program failed after a major investment in the program
- A habit of virtually always "refining" or "tailoring" an approach before accepting it
- A group that is anxious to try something different or be the model for other departments or divisions
- A desire to be exemplary and follow the leading edge business practices (e.g., ISO certification, the Malcolm Baldridge award, continuous improvement, or six-sigma)
- Limited resources—whether it be for training or in terms of flexible project managers.

To prove the benefits of MERITT to yourself and your company take a noncritical project and tackle it with and without MERITT. Alternatively, use a recently completed project and apply MERITT retrospectively or use a project that has been put on hold and apply MERITT to the remaining development stages.

After demonstrating success through limited scale implementation and gathering enough company-specific evidence showing the value of MERITT merged into the existing process development process, the champion(s) will eventually need to take the story to top management. At this point, the MERITT-influenced development process can be turned into an overall policy or a corporate-wide approach.

### Command and Control Cultures

In these companies an immediate and total change will be necessary to reinforce the message that, "This is how things are going to be done, no questions asked." This can be true in countries where employees always look to their bosses for direction or in companies or countries where there are rigid rules and guidelines for all business processes. Full-scale adoption can also work when many processes are being changed as a result of a merger, another major event like a turnaround, the occurrence of a catastrophic event or traumatic experience, or even a new training program. Companies that really push back at any kind of change to the status quo will also occasionally be more likely to adopt a rapid and mandatory change than an evolutionary one, especially if their future viability is on the line.

### Guerilla MERITT

In yet other companies, there may be a small cadre of believers, but no initial top-down support. In these cases, it will be important to find a process development team where the benefits can be demonstrated. The process development team might be penetrated by the small set of "believers" via:

- A supportive project manager, probably with strong personal relationships to one or more of the believers,
- An engineer on the team who has an EHS background or good experience with EHS tools,

- A very experienced team member who can personally bring IS, P2, and GC perspectives to the team, or
- A very inexperienced project manager who is looking for guidance and a different pathway to success.

In this scenario it will be important to document the process used to follow MERITT as well as all the benefits and lessons learned. This information should be used to refine the approach and then additional pilot opportunities sought. Once a compelling story can be told for several different projects (to avoid the appearance of a one-shot wonder), the believers need to start a careful communication campaign that reaches out to senior management as well as those with various project responsibilities. This will provide both a push and a pull on the organization and should help ensure greater support and buy in. Parallel efforts to develop tools and offer training will help ensure that potential obstacles are minimized before someone has the opportunity to say that they are insurmountable.

## 7.2. Integration with Product Development Process

The underlying *product* development process drives most *process* development processes and the addition of MERITT must not impede any ultimate commercialization. At the same time, MERITT offers value long before the decision is made to pursue commercialization and care must be taken to push MERITT as far back into the product development cycle as possible.

Demonstrating the potential to actually help or hasten the time to commercialization will ensure much more rapid implementation of MERITT. For instance, Monsanto's DSIDA example (described in Chapter 2) is one case where the concurrent consideration of a number of EHS perspectives led to a solution that was implemented quickly—24 months from the initiation of the project to starting operations.

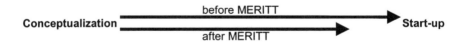

MERITT should be applied only to the level of detail needed to realize the desired benefits. It should not be viewed as an inflexible set of
requirements, as this may often be nonproductive. Just as the process
development cycle varies depending on the magnitude of the project and
its nature, so should the elements of MERITT that are applied. The fundamental principles will remain constant, but the resource components
and the implementation elements will be different for each situation.

The starting point for the initial implementation of MERITT is decid-
ing what the enhanced process development process is supposed to
achieve. This may involve both primary and secondary requirements.
Primary requirements might be to ensure that EHS issues are addressed
concurrently and early so that the process is ready for startup on schedule and performance is sustained during operations. A secondary
requirement could be to ensure that specific regulatory requirements
(such as a PHA study) are completed at appropriate times.

After the objectives have been identified, the next steps involve
designing the critical "tie lines" between the process development process and MERITT. These will include the criteria to be used at various
stage gates (see Chapter 4). The expertise needed for each stage must
also be delivered, whether in person or through tools and training. This
may require changing the staffing mix for the process development
team or developing and/or training staff on various tools that effectively
bring EHS knowledge into the evaluation and decision-making process.

Once the basic flow of the MERITT-influenced development process
has been outlined and thoroughly reviewed, implementation can begin,
leading to the ultimate adoption of MERITT as a core competency.
Interim arrangements (e.g., additional internal or external resources)
may be required until certain tools or training efforts are complete, and
until the MERITT approach becomes more natural.

MERITT implementation will be more successful if the approach has
been reviewed by representatives of those groups who will be affected
by it, i.e., the stakeholders. During these reviews, most project managers and team members will share some concerns about merging MERITT
into the existing process development process. Some potential concerns
are described below along with ideas on ways to defuse them.

- *Span of control.* Does someone other than the project manager
  have control?

> The basic role of the project manager (and the overall team) will not change and this should be emphasized in communications. The day-to-day process will still be driven by the staff involved, but those staff may change slightly and may involve other groups besides engineering and chemists. The role of the manager will still be to monitor performance and make sure that appropriate resources are available—albeit a wider range of resources and tools.

• *Development process costs.* Will these increase significantly? This includes not only the EHS staff resources that may be involved at various stages, but also any additional time needed from the typical team members.

> Start with external examples demonstrating the benefits from using the MERITT approach and some of the pitfalls associated with using uncoordinated, sequential individual E, H, and S reviews. If you have any company or industry experience with some of these pitfalls, these should be incorporated from the beginning. As internal successes occur, they should be promoted and included in training packages.

• *No improvement in EHS performance (or reduction in cost) will be achieved.* Project managers and others facing a change in a tried-and-true process often express the concern that the change will not achieve its objectives.

> Almost by definition, integrated processes have more and bigger opportunities for enhanced and sustained performance than can be gained from the collection of individual programs they replace. Look to internal experience with other integration or change activities within your company; draw on the successes associated with implementing these other changes. If the success was limited, investigate what went wrong before and try to address the concerns in a way that avoids past problems.

• *Lack of goals or objectives.* Are there specific goals and objectives for MERITT? How far should the integrated process go?

> If no integrated EHS or MERITT goals are developed, the development process, and therefore the overall project, can be slowed down significantly. This can also lead to the adoption of default goals from other sources that do not really apply.

- *Doing more work.* If certain groups have not historically been involved there can be questions of how they can possibly do "more" work.

> Work on demonstrating the change in a broader set of EHS activities so that the benefits that may be offered later on are recognized. There should ultimately be a savings in EHS work, not an increase. The use of tools instead of EHS staff in various stages can also help with workload issues. However, the utilization rate for EHS resources and the distribution of EHS staff by skill area could potentially change.

- *Getting the right expertise at the right time.* There may be difficulty in convincing people that new faces (or perspectives via tools) add value.

> You will need to collect and communicate your own examples of success as they occur to support the use of the right people (or tools) at the right time. There can also be significant obstacles to identifying the right people, let alone the right people who also have the personalities to fit in with the teams. When such people are identified they may have to be made more available to facilitate the adoption of MERITT. It will be very important to provide ongoing information on how to access the necessary expertise.

## 7.3. Overcoming Behavioral and Cultural Barriers

Cultural barriers can be as basic as different groups being housed in different buildings or even different towns. In other cases the barriers may be associated with the biases that people have regarding the skills and competencies of other groups or individuals.

Cultural problems often arise due to misalignment between new requirements and factors that drive day-to-day behavior—also known as *the unwritten rules*. Behavior is generally influenced by a combination of three different factors:

- *Motivators*—drivers for particular behaviors
- *Triggers*—events that can initiate the whole process of motivation
- *Enablers*—people or systems that can provide/enhance the motivators

The goal is to understand the motivators, find and use the appropriate triggers, and obtain the support of the enablers in order to change behaviors. An *unwritten-rules* assessment may be helpful if there is major opposition or difficulty in implementing MERITT, but if it is truly merged throughout the existing process development process, the opposition is unlikely to be that difficult to surmount.

Identifying the benefits of the revised process development process for different team members can help motivate acceptance or at least tolerance of the change and help them get past the perspective of "we do this well already."

Past experience in implementing other corporate initiatives can help you determine which benefits are most likely to be compelling or applicable to particular individuals and groups and address the "what's in it for me" question. Such benefits might include:

- *Less time spent on EHS issues*. Although the coverage of EHS issues is not what is desired in current process development processes, a lot of time is still spent on these issues. And despite the time spent, project managers may still worry about the things they don't know (due in part to little EHS training) that may cause problems in the future. With MERITT integrated into the process development process, project managers should find themselves spending less time worrying about unknown EHS issues and more time using EHS information to make better decisions.
- *Fewer processes to manage and results to interpret*. Once MERITT becomes routine, the total number of separate evaluations embedded within the development process should decrease, along with the quantity of independent results. The path to using the information in decision making should be much more straightforward.

- *Better measurement of process performance.* Integrated EHS metrics and the use of EHS metrics concurrently with other performance measures should provide a much better picture of the predicted performance for a new process or for different alternatives that are under consideration.
- *Clearer roles and responsibilities.* Chances are that different team members are trying to informally introduce some consideration of EHS at different points in the process, but their contributions are limited by their own past experience with these issues. They want to help, but they don't want to be considered experts in areas that they really aren't.

It is important to obtain top management commitment to MERITT as discussed in Chapter 4. The successful implementation of MERITT as part of the existing process development process will be both easier and more successful with strong management commitment. Of course, such commitment will be forthcoming only if it is clear that MERITT is value-added. Champions in top management or with top management's ear are essential and will most likely have the following characteristics:

- Impatience with the inefficiency, ineffectiveness, and independence (and cost) of current EHS evaluations,
- Dissatisfaction with existing EHS measures and the lack of business considerations, and
- Reputation for embracing managed change, particularly in established processes.

The expectations for the top management champion must be consistent with the company's standard operating style. Senior managers accustomed to and known for delegating decisions will want to maintain that style and may not be effective champions.

Some specific potential barriers to look out for throughout an organization include:

- *Accepting the Ideas of Others.* As with modifying the composition of the team, there may be barriers in getting their voices heard. This can be overcome by developing believers and coaches among existing team members, cross training highly respected individuals, ensuring that only the best people get put on the first few teams, having open-minded senior managers facilitate team meetings, working with project managers to ensure their buy-in and support of MERITT, publicizing success stories, and rewarding new ideas and solutions on a team basis.

- *Using New Tools.* Where companies are used to using certain tools that do not readily lend themselves to being made MERITT friendly, it may be necessary to provide training in new tools, use two tools in parallel until the new ones are accepted, or modify the tools in-house so that they become a "company" tool. Discussions with respected peer companies that use the new tool or a variant on it may be a quick way to create converts.

## 7.4. Metrics

Measuring ongoing performance is a vital part of any good process or system, and it is often stated that it is impossible to manage effectively without measurement. For MERITT, metrics should address both the predicted EHS performance for the new or upgraded process design (at different stages of the development process) and the efficiency of the MERITT-influenced development process so as to measure the cost/benefit of the modifications to the development process.

In the former case, the metrics are likely to be used to make comparisons of different alternatives or to assess new process designs against existing ones and to measure actual performance of the MERITT-influenced process as designed. In the latter case, measures are needed that quantify the benefits accruing from the enhanced development process, such as the time to commercialization or startup, or the total quantity of EHS resources used on a particular project.

Established performance measures have a role in the transition to a MERITT-influenced approach, even if they will not be maintained long term—they can provide before and after comparisons and they can

Gathering data for a few projects that do not use MERITT will be very useful in establishing benchmarks for future projects to be compared against and to develop "success stories" that will help to motivate some of those who may oppose merging MERITT into existing processes. Tracking the amount of rework associated with these non-MERITT projects can further support the value of MERITT.

provide some indication of EHS performance to decision makers until the decision makers fully understand and trust any new metrics that are put in place. Also, while the objectives or requirements for a particular process development exercise will influence the selection of metrics, there should be some metrics that are always examined in order to allow comparisons across projects and against overall corporate goals.

Metrics allow the establishment of the benchmarks for answering the question "Have we achieved our objective?" Metrics can be leading, real-time, or lagging indicators of performance. Many EHS metrics are lagging indicators as these are generally easy to establish, track, and communicate. For example, the lost time incident rate (LTIR) provides statistics on past and current safety performance, but is less valuable in predicting future performance.

In the process safety practice area, risk-based metrics such as criti-

> For metrics to be useful to the application of MERITT, they must be **leading indicators** of future EHS performance, otherwise, they provide little utility to the process development team.

cal event frequencies or frequency versus number of impacts (typically injuries or fatalities) plots (also known as F-N curves or risk profiles) are employed to help predict and assess future performance.

In general, the use of overarching MERITT metrics is not very common in EHS practice. One reason is that there are few metrics that truly reflect integrated EHS values. For acute and some chronic risks, the application of metrics such as individual and societal risk tolerability criteria in combination with quantitative risk assessments (QRA) are sometimes utilized for highly hazardous activities, to decide when enough risk mitigation and management have been provided. To the extent that cost is the ultimate discerning metric, tools such as LCA and TCA can support financial metrics. Identifying metrics that measure the true extent to which MERITT has been achieved are probably more critical to the ultimate success of MERITT than the methodology itself. Having metrics will not only provide an incentive to use the tools, but may also indicate which tools are most important for different circumstances.

CWRT has coordinated an industry-led effort to develop and test sustainability metrics useful for tracking operational improvements.

Metrics have been developed for mass and energy intensity, water usage, pollutants, and health- and ecotoxicity. For mass and energy, core metrics as well as complementary metrics have been found valuable. Some of these metrics have also been tested on a number of the major extant chemical processes. The sustainability metrics are available on AIChE's web site at www.aiche.org/cwrt and could provide a starting point for the development of company specific MERITT metrics.

Other possible measures for MERITT include:

- *Quantities and costs of emissions to air, water, and land; liquid and solid effluents; and hazardous material releases.* Generally these measures reflect only the particular materials or usage rates of concern (e.g., steam releases may not be included). Historically they have focused on the quantities and not the costs, but both aspects are critical.
- *Mass and energy balances or requirements, with costs.* These are basically traditional metrics that are likely to be used in a company's existing process development process, but they also provide insight into MERITT concerns. These measures can provide continuity between the existing approach and the revised process development process that reflects the MERITT approach. The balances have often been used by the engineers and the costs by the financial specialists and senior management, but the costs should be considered early on by all parties. (Also see CWRT metrics.)
- *Ecological impact of operations.* This has traditionally been difficult to measure consistently, but some of the tools such as TCA are finding ways to develop meaningful metrics for this area, including financial impacts. (Also see CWRT metrics.)
- *Hazard assessment findings or results.* These predictive results might be generated as part of hazard assessments such as a Hazard and Operability (HAZOP) study, a what-if/checklist, or other techniques as described in Chapter 5. Such studies indicate the types of hazards of concern in addition to the types and effectiveness of the controls in place.
- *Risk assessment results.* Such results provide an overall estimate of the EHS (or just E or H or S) risks associated with a process in whole or part. By applying a logical and systematic evaluation approach (be it simple or sophisticated), these studies support consistent and cost-effective decisions on the need for additional

controls or changes. By assessing different types and sources of risk, such studies are very useful in identifying (and ultimately resolving) trade-offs between different options.

• *Hours (and costs) dedicated to EHS evaluations and risk management.* The concern here is with the total expenditure of resources over the life cycle of the process, not just those needed for the development process or, later on, those needed to provide risk management services on a day-to-day basis. (This requires good benchmarking data before MERITT is introduced and is intended to cover the long-term costs of all EHS resources, whether they are provided on site or by a corporate group.)

When it comes to metrics, the general rule is "the fewer, the better," but this assumes that the measures are good ones and that they can be used and tracked without undue difficulty. Too many metrics leads to confusion as to what one has really learned about a process, as well as to wasted resources. The use of too many independent metrics could also promote the view that EHS really is not integrated through the application of MERITT.

For many of the metrics that can be applied in a predictive fashion, there may need to be thresholds defined to provide guidance to the design team as to what the goals should be in different areas. Such goals or targets will be very specific to a given company.

> The "best" metrics for MERITT may well be new ones that combine TCA and sustainability metrics to support the choice of alternatives considering both economic and environmental impacts.

Resource-based measures can provide a much better way to manage headcount based on the knowledge of actual resource needs, rather than just undergoing arbitrary reductions when budgets are tightened.

## 7.5. Addressing Industry-Specific Issues

Specific industries are likely to have: alternate definitions of stages (or different milestones), stage compression or expansion, different evaluation criteria, varying levels of EHS expertise available internally, and different preferred EHS tools. All these differences must be acknowledged and addressed as part of merging MERITT into the existing process development process or the concept of MERITT may not be seen as fitting and may not be accepted. Industry specific tailoring will likely mean:

- Developing a specific training package that tailors MERITT for in-house use.
- Establishing a set of stage, expertise, and tool definitions that are specific to the industry or company. These should be widely circulated internally and/or on an industry web site for "validation" as equally useful and good (to those presented herein).
- Making presentations for industry conferences and possibly forming industry task forces or roundtables to aid in implementation and broader adoption (good for joint tool and metric development).

The definition of a successful MERITT program may also vary by industry, as their needs may be different.

## 7.6. Applying to Small Companies

The same basic concepts and approaches can be applied at small companies, but it may be more difficult to get the needed expertise and perspectives if they are not captured within a tool. The availability of all process development resources may be very restricted, necessitating very simple and rapid tools or other approaches. There may not even be time for formal training for project managers or other team members.

Under some circumstances it may be worthwhile to get outside help (from consultants, parent companies, industry associations, or via other alliances) to ensure that MERITT is developed and incorporated into existing development processes in a cost-efficient and effective manner. In such situations, the goal should be to use the outside resources to get the new approach jump-started, but not long term except when very complicated issues arise.

## **7.7. Evaluating Licenses**

A license, by its nature, requires a is a well-defined result. This often makes for inflexibility due to an inability to change design parameters and conditions. Therefore the use of MERITT in such situations may not be to enhance or optimize design but rather to evaluating competing or alternative technologies and approaches to see which one is initially more MERITT-like and which one might allow MERITT-driven improvements in the future.

The criteria or stage-gate requirements may help define a unified set of MERITT criteria or specifications that can be passed on to technology developers to ensure that they optimize their future technology developments.

# Appendix A

# Additional Tools with Applicability to MERITT

This appendix lists references to additional tools that are applicable to MERITT. Specific areas of tool applicability are given in parentheses after some of the references. Comments further define the applicability of particular sources.

| Title | Comments |
|---|---|
| Benson, R. S. and J. W. Ponton. "Process MIniaturization—A Route to Total Environmental Acceptability?" *Trans. IChemE,* 71, Part A, 160–168, March 1993. | |
| Berger, Scott. The Pollution Prevention Hierarchy as an R&D Management Tool (PP via process and product modifications, 1994). | |
| Cave, S. R. and D. W. Edwards. "Chemical Process Route Selection Based on Assessment of Inherent Environmental Hazard" *Computers Chem. Eng.,* 21, S965–S970, 1997. | |
| Dunn, R.F., et. al. "Application of Process Integration Technology in the Chemical Process Industries." Presented at AIChE 1999 Spring Meeting, Houston, March 1999 (russel.f.dunn@solutia.com). *(Option Generation, Evaluation/Decision Support)* | Process integration design methodologies for waste reduction (mass and energy) utilizing heat and mass exchange networks. |
| Fiksel, Joseph. "Methods for Assessing and Improving Environmental Performance," Chapter 9 of *Design for Environment.* | |

| Title | Comments |
|---|---|
| Golonka, K. A. and D. J. Brennan. "Application of life cycle assessment to process selection for pollutant treatment." *Trans IChemE*, 74 B, 105–119, 1996. *(Evaluation/Decision Support)* | |
| Gupta, J. P. and B. S. Babu. "A New Hazardous Waste Index." *J. Haz. Mat.*, A67, 1–7, 1999. *(Evaluation/Decision Support)* | |
| ICI, Environmental Burden: The ICI Approach (A new method to evaluate the potential environmental impact of waste and emissions), 1997. *(Evaluation/Decision Support)* | Comprehensive index to estimate the potential environmental impact of gaseous and aquatic emissions |
| Kollman, C. J. "Achieving Environmental Excellence in Process Development." Session 20, Pollution Prevention. HAZMAT International '93 Environmental Management and Technology Conference, June 9–11, 1993, Atlantic City, N.J. | |
| Johnson, R.W., S. D. Unwin, and T. I. McSweeney. "Inherent Safety: How to Measure It and Why We Need It." International Conference and Workshop on PSM and Inherently Safer Processes, October 8–11, 1996, Orlando, FL. New York: AIChE, 118–127. *(Evaluation/Decision Support)* | Presents three measures of IS based on contained energy, affected area, and potential impact. |
| Lawrence, D. and D. W. Edwards. "Inherent Safety Assessment of Chemical Process Routes by Expert Judgment." *The 1994 IChemE Research Event*, 886–888. *(Evaluation/Decision Support)* | |
| Lawrence, D., D. W. Edwards, and A. G. Rushton. "Quantifying Inherent Safety." *Proc. Inst. Mech. Eng.*, 4, 1–8, 1993. | |
| Mak, C.P., H. Muhle, and R. Achini. "Integrated Solutions to Environmental Protection in Process R&D" *Chimia*, 51, 184–188, 1997. *(Option Generation)* | Very good pointers from the chemist's point of view, and for addressing P2 and IS issues early on (even though not called as such). |
| Wood, M. and A. Green. "A Methodological Approach to Process Intensification." *IChemE Symposium Series*, 144, 405–416, 1998. *(Option Generation)* | Valid attempt to structure process intensification as a methodology. Worthwhile reading, both for the methodology itself, and for good examples of lesser-known options for inherently safer design. |

# Appendix B

# Overview of the INSET Tools and Their Aims

| Tool | Name and Aim |
|------|--------------|
| A.1 | *Detailed constraints analysis*—to define the limitations and boundaries of the project. |
| A.2 | *Detailed objectives analysis*—to define the aims and goals of the project. |
| B | *Process option generation (incl. Process waste minimization guide)*—to rigorously challenge route and process alternatives in order to obtain a more ISHE process. |
| C | *Preliminary chemistry route options record*—to consistently present all the proposed chemical route alternatives. |
| D | *Preliminary chemistry route rapid ISHE evaluation method*—to provide a rapid assessment procedure to determine the most viable chemical route alternatives. |
| E | *Preliminary chemistry route detailed ISHE evaluation method*—to evaluate the chemical route alternatives with respect to the constraints and objectives that define the process. |
| F | *Chemistry route block diagram record*—to give an overview of the process involved for each alternative. |
| G | *Chemical hazards classification method*—to provide a simple and easy-to-apply means of classifying materials in terms of their hazardous properties. |
| H | *Record of foreseeable hazards*—to identify possible hazards caused by the desired or an undesired reaction, and record these. |
| I.1 | *Fire and explosion hazards index*—to provide a means of comparing route alternatives on the basis of the potential for fire or explosion. |
| I.2 | *Acute toxic hazards index*—to provide a means of comparing route alternatives on the basis of the acute toxic hazards. |
| I.3 | *Health hazards index*—to provide a means of comparing route alternatives on the basis of their health hazard performance. |

| Tool | Name and Aim |
|------|--------------|
| I.4 | *Acute environmental incident index*—to provide a means of comparing route alternatives on the basis of the potential to cause acute environmental incidents. |
| I.5 | *Transport hazards index*—to provide a means of comparing process route alternatives on the basis of their transport hazards (accidental releases of material during transport off-site). |
| I.6 | *Gaseous emissions index*—to provide a means of comparing process conditions and plant alternatives on the basis of the potential to cause routine/daily impact on the environment. |
| I.7 | *Aqueous emissions index*—to provide a means of comparing process conditions and plant alternatives on the basis of the potential to cause routine/daily impact on the environment. |
| I.8 | *Solid wastes index*—to provide a means of comparing process conditions and plant alternatives on the basis of the potential to cause routine/daily impact on the environment. |
| I.9 | *Energy consumption index*—to provide a means of comparing process conditions and plant alternatives on the basis of the potential energy usage and the resultant effect on the global environment. |
| I.10 | *Reaction hazards index*—to provide a means of comparing process conditions and plant alternatives on the basis of the potential for runaway reactions. |
| I.11 | *Process complexity index*—to provide a means of comparing process options on the basis of their likely complexity, hence difficulty to control and prevent errors. |
| J | *Multiattribute ISHE comparative evaluation*—to provide a means of evaluating and comparing the ISHE performance of various aspects of the route alternatives as a means to eliminate the more unfavorable process options. |
| K | *Rapid ISHE screening method*—to rapidly assess each route alternative with respect to its ISHE performance, as a fast-track alternative approach to Stage II. |
| L | *Chemical reaction reactivity–stability evaluation*—to identify any chemical process that may have runaway potential or in which other hazardous situations may occur due to chemical reactions. |
| M | *Process SHE analysis/process hazards analysis and ranking*—to provide a simple method to identify and rank any hazards in the proposed process. |
| N | *Equipment inventory functional analysis method*—to provide an understanding of why inventory is required at a plant, leading to the generation of ideas on how it might be minimized. |
| O | *Equipment simplification guide*—to challenge the need for valves, instruments, flanges, and other pipework or equipment fittings that can increase the complexity of the plant and maintenance requirements. |

| Tool | Name and Aim |
|------|--------------|
| P | *Hazards range assessment for gaseous releases*—to provide engineers with an easy-to-look-up indication of the magnitude of major accident hazards based on either the process inventory or the size of typical leak sites. |
| Q | *Siting and plant layout assessment*—to challenge the basis of the plant layout at the early stages of its development, in order to see how changes to the layout could improve segregation and make the layout more inherently SHE. |
| R | *Designing for operation*—to provide a simple checklist for those involved in the detailed design of plant to prompt them to consider ways in which to make the plant easier to operate and maintain. |

For information regarding access to the INSET toolkit, contact the project manager, David Mansfield (dave.mansfield@ aeat.co.uk).

## Appendix C

# The Business Case for Managing Process Safety

## Managing Process Safety Is Good Business

The Center for Chemical Process Safety (CCPS) of the American Institute of Chemical Engineers developed a business case for process safety management. This landmark study, based on input from CCPS member companies and other sources, conclusively demonstrates that managing process safety provides major quantitative and qualitative business benefits by providing companies the freedom of self-determination, avoiding the major costs and injuries associated with accidents, establishing corporate responsibility, earning the respect of the local community, and creating business value.

Self-Determination

Value

PROCESS SAFETY

Loss Avoidance

Corporate Responsibility

## Self-Determination

Managing process safety gives a company the freedom to manage its business and grow profitably, while still satisfying all stakeholders. Managing process safety helps assure that a company retains its "license to operate and expand" with the full support of the public and the local community. Companies that avoid major accidents maintain good relations with the regulators and the community and are often able to obtain expedited approvals for permits to expand and in some cases the approval to site new facilities. This can be critical to the creation of new products necessary for the company to compete effectively. Major accidents have catalyzed the promulgation and enforcement of regulations, for example, Bhopal, and more recently Tosco in California and Napp Technologies in New Jersey. Every company in the same industry suffers after a major accident. Employees, managers, officers and board members can have their jobs "on the line" by not having an effective process safety program. A company's future can be severely impacted by a major incident.

> Managing process safety maintains a company's freedom of self-determination.

## Avoiding Major Losses

This is another benefit realized by companies that have implemented effective process safety programs. Accident prevention avoids:

- Loss of lives of and injuries to employees, contractors, and the public
- Property damage from a major accident—an industry-average cost of $80 million per accident
- Business interruption losses from a major accident—four times property damage, plus the almost certain loss of market share, at least for the short term, until the company's reputation is restored
- Regulatory fines for major accidents which routinely exceed $1 million plus litigation costs that may be five times higher
- The cost to investigate the accident and implement corrective actions

A majority of the companies that participated in this study achieved significant reductions in injury rates from their process safety program. One company achieved a 50% reduction in injuries and fatalities resulting from major accidents (compared to overall industry averages)—thus saving over $5 million per year and an additional savings of $3 million per year in workers compensation costs. Many companies feel that their

Process Safety programs have helped them avoid a major accident following downsizing in their organization.

> Managing process safety protects a company from major losses of life and property.

## Corporate Responsibility

Corporate responsibility means doing the right things as responsible members of the community—compliance with regulations, conformance with internal and industry standards, and maintenance of a safe work environment—an obligation to employees, the community and the company. Employees in a responsible company have higher morale and loyalty; this makes it easier to retain and hire new employees and enhances employee relationships. The company can more readily attract staff for leadership positions and directors for the company's board. Today communities have "zero tolerance" for companies that have incidents with offsite impact. A responsible company uses process safety to reduce the likelihood of experiencing a major accident or adverse regulatory inspection, and to enhance its corporate image. A reputation for being a safe company increases the value of the company brand and may result in higher sales and market share. The lower risk perceived by the investment community could increase a responsible company's share price.

> Managing process safety protects a company's reputation and shareholder value.

## Value

Value is created by doing the right things right. Managing process safety well allows a company to achieve substantial increases in revenues and reductions in cost. The value created can be substantial as shown in the following data supplied by large chemical and petroleum companies that participated in this study:

- Up to a 5% increase in productivity, primarily due to increased reliability of equipment, resulting in increased revenues of $50 million
- Up to a 3% reduction in production cost resulting in a savings of $30 million
- Up to a 5% reduction in maintenance costs resulting in savings of $50 million
- Up to a 1% reduction in capital budget resulting in savings of $12 million
- Up to a 20% reduction in casualty insurance cost resulting in savings of $6 million

For smaller companies, implementation of process safety may be a prerequisite to doing business with responsible companies due to product stewardship requirements.

Managing process safety creates company and shareholder value.

This study has demonstrated that managing process safety well creates both financial and nonfinancial business value. Many leading companies implemented process safety programs company wide before there were quantitative data to support the business case and have already gained this competitive advantage. Good management of process safety can also help to meet a wide range of corporate obligations, avoiding a multiplicity of required programs by combining them into one (regulations, ISO, TQM, sustainability, responsible practices, etc.).

Realize the full business benefits of managing process safety well in your company.

# Subject Index